Gumshoe Geography

Gumshoe Geography

Exploring the Cultural,
Physical, Sociological,
and Biological
Characteristics of Our Planet

Richard S. Jones

Zephyr Press ®

REACHING THEIR HIGHEST POTENTIAL

Zephyr Press
Tucson, Arizona

Gumshoe Geography
Exploring the Cultural, Physical, Sociological, and Biological Characteristics of Our Planet

Grades 6–12

© 1996 by Zephyr Press
Printed in the United States of America

ISBN 1-56976-028-4

Editors: Stacey Lynn and Stacey Shropshire
Cover design: Nancy Taylor
Design and production: Daniel Miedaner

Zephyr Press
P.O. Box 66006
Tucson, Arizona 85728-6006

CURR
G
73
.J66
1996

Library of Congress Cataloging-in-Publication Data are available.

Contents

Teacher's
Instructions

Introduction

Learning Styles and Intelligences

We know that learners learn in a variety of ways, that there are several learning styles, and that students have multiple intelligences. Throughout the student activities, students are asked to report on a variety of questions and topics. *Report* can be understood in many ways. I use the term to include a variety of methods of "reporting." To take into account students' learning styles and intelligences, you might have them report through any of the following means or in other ways that you choose:

Murals	Videotapes
Skits	Riddles
Songs, raps, or rhymes	Collages
Poems or stories	Scavenger hunts
Models	Maps
Games	Paper folding
Editorials	Mobiles
Comic strips	Crossword puzzles

Guide to Data Report Form

Use the chart (page xxxi) and instructions at your discretion. You may choose to use only one or two of the boxes in the chart or you may choose to use the entire chart for an exhaustive study. Please keep the following guidelines in mind:

1. No two schools or classrooms will have the same information sources available. Have students try as hard as they can to find appropriate sources and search out the information needed for the problems.

2. It is important for students to be frustrated sometimes and not be able to find all that is required of them. This frustration helps them learn to help and depend on their peers in cooperative situations, to make educated guesses, and to speculate based on previous learning. It increases their curiosity and appetite for the information they are seeking.

3. When students have exhausted the available resources, encourage them to speculate on areas such as food and clothing. Also encourage them to use alternative sources of information, such as personal interviews of people with firsthand experience of the places they are researching and computer resources. CD-ROM technology, other computer databases, the Internet, and other on-line resources open the world of information to just about anyone.

4. Whether the students find information from a direct source or end up speculating, they have to understand that they are ultimately responsible for the information. They should be ready to justify each and every response on their charts.

Teacher's Goals and Objectives
Planning Work Sheet

Map Activity Title: _____

Assignment date: _____/_____/_____   Due date: _____/_____/_____

Students are working _____ individually _____ small groups _____full class

Objectives

Subject Areas/Lesson Coordination: _____

Expected Learning Outcome: _____

_____

Parts of Data Report Form Required

____ Coordinates	____ Destination	____ Mode of Travel
____ Terrain	____ [icon]	____ [compass icon]
____ Distance	____ Arr./Dep.	____ Dietary Needs
____ Clothing	____ Special Notes	____

Will *Your Assignment* be required? _____

Will *Special Instructions* be required? _____

Who will do *Special Instructions* and *Your Assignment* and how will they be credited?

_____

_____

_____

Special Grading Notes: _____

_____

_____

Student Training Manual

Introduction

This is the age of computers, satellite communication, high-speed travel, and information, information, information. International trade barriers are being broken down. Goods are being made in some countries and assembled in others. The world is becoming smaller; access to the world community is becoming easier and easier.

Because of these changes, geographical knowledge is becoming more and more a key to success in the world and should be an important part of any serious student's academic preparation. There are a few basic skills that everyone needs to master to be able to understand geography and use its tools for the greatest benefit. This training manual offers instruction and practice in those skills. You can make the training personal and customize the lessons to satisfy your own curiosity. If a lesson seems too easy, make sure you understand the skill that it is teaching. Have fun and open the doors to the world!

The Committee

We in The Committee have been watching you, and we have determined that you are qualified to enter the training program to become one of our members. Your teacher will be your supervisor and will help train you. When your supervisor feels that you are ready for service, you will be notified and assigned a mission within The Committee.

The training program is designed to teach you the few important skills that you will need to see the world and then report what you have seen and learned. Follow directions and practice hard. We are looking forward to having you join us in The Committee.

Training Program Form

Name:_____

Grade: _____ Teacher: _____

Check off the parts of the training that you have completed and passed:

___	Pronunciation	__	Direction
___	Hemispheres	__	Latitude and longitude
___	Using grids, parallel	__	Using grids, curved
___	Map scale and distance	__	Reading topical maps

Gumshoe Geography © 1996 Zephyr Press, Tucson, AZ

Direction

We use two types of directions that have to do with the geography of Earth. Both use the words *north, south, east,* and *west.* One describes where you are and the other describes where you are going.

The "where you are" direction describes what quarter of the globe you're on. (We'll explain this concept more later.) The Earth spins on an imaginary center line called an *axis.* When you spin a basketball on your finger, your finger is the axis. The ends of the Earth's axis act like a magnet, except that the two poles are called *North* and *South* instead of positive and negative. If you are closer to the North Pole, you are in the North and the same goes for the South. That's easy.

East and West are a little harder. For now, just remember that if you are in North, South, or Central America, you are in the West, and if you are anywhere else, you are in the East of the Earth. So you are in two places at the same time: North and West, North and East, South and West, or South and East. This direction is the "where you are" direction.

The "where you are going" direction relates to which section of the globe you are facing when you decide to move. You can always tell which way you are going by using a compass or by using the sun. If you are in the North, the sun at noon will be in the South. The opposite is true if you are in the South. East and west are easier "where you are going" directions than they are "where you are" directions.

If you are facing north, west is always to your left and east is always to your right. You could travel west forever and even though you could be in the East or the North or the West, you would still be traveling west.

As with the "where you are" direction, you can use two of the directions at the same time, one of each: northwest, southeast, and so on. But you can't travel northsouth or eastwest.

Most maps have a symbol called a compass rose that tells how the map is laid out in terms of these directions. The compass rose looks like the symbol at the right. Although it does not always point toward the top of the paper, it always points north.

Gumshoe Geography © 1996 Zephyr Press, Tucson, AZ

Hemispheres

The world is a big ball spinning in space. We won't get into the details of that right now; just be satisfied to picture your home as a ball spinning in space!

A long time ago, to satisfy a human need to communicate where people were and where they were going, our ancestor—we'll call him Don Geo—invented geography, a system of naming and locating places on the globe. Your personal address and everybody else's refers to geographical locations. Let's look at a typical address:

> Don Geo
> 115 South Street
> Westerville, North Dakota

Now let's extend this address to see more exactly where Don lives.

> Don Geo
> 115 South Street
> Westerville, North Dakota
> United States of America
> North America
> Western and Northern Hemispheres
> The World

As the address gets longer, the divisions of the world get more general and larger. The largest and most general division of the world is called a *hemisphere,* which means half of a sphere. The globe is invisibly cut in half either top to bottom or side to side. This division is great, but something is missing; it isn't specific enough. Don saw this problem and decided that the world could be divided in half in two ways:

left and right,
or East and West.

and top and bottom,
or North and South

Don was feeling smart and smug when a new problem popped into his brain. North and South were easy enough to figure out—one was up and one was down; the compass pointed north and that was that. But where did East begin and where did West end, and how do you tell the difference?

Don got together with a group of his friends and they began to think. They decided that there would have to be a starting place, and after much discussion they decided that it would be a place called Greenwich, England, which is near London. Everything east of this place to the exact other side of Earth would be East, and everything west to the exact other side of Earth would be West.

Of course, dividing the world in half in two ways really divides it into quarters, but we still refer to halves or hemispheres. Therefore, in the United States, we live in the Northern and the Western hemispheres.

Longitude and Latitude

Let's look at some examples of grids:

⊕ When two people play checkers, they use a board with squares on it. This layout of squares is called a *grid*. The grid lets the players know where they are, how far they've come, and how far they are going.

⊕ When blueberry pickers head out into a field for the harvest, they use string to divide the field into large squares so that each picker has his or her own territory to pick completely clean.

⊕ When city planners lay out a new city or a new part of an older city, they often try to lay the transportation corridors out in a grid so that all parts of the area are easy to find and get to.

There are many other examples of people using grids to help them do things. Can you think of any? Name two!

Our hero, Don, still had a problem with showing others where he lived using a map. Finally the idea of a grid popped into his head, and he decided to divide his map into columns and rows of equal sizes. He numbered the columns and lettered the rows. Particular locations could be found by crossing a column with a row to find a square in which the place appeared.

⊕ Look at a map in an atlas and find this type of grid. See how it works, and see if you can understand how Don feels about this grid location system.

This grid provided an easy way to locate places on any given map, but there was a problem: it wasn't universal or very accurate. Don's city might be found in the M-8 square on one map and in Q-17 on another map, and his friends had to search the square to find the exact location. Don needed something more accurate.

Don thought that his idea of hemispheres was a good one and that if he took it another step, the problem might be solved. He knew that a circle is divided into 360 sections called degrees, so he figured that since the Earth is round, he would start with these divisions. When he had finished, the Earth was divided into a cross grid using markers, 360° from North to South and 360° from East to West.

To make the grid even more accurate, Don used the system of dividing each degree into 60 minutes—in both directions. A minute of distance, at the equator, comes out to about 1.3 miles. Each minute is also divided into 60 "seconds." A second of distance, at the equator, is equal to about 118 feet. This grid made locations on any map consistent.

The two groups of lines that divide the Earth into the imaginary grid are quite different from each other, and they have different functions. We will talk about them separately, and the best way to understand is to have a globe to look at as you read this lesson. Try reading each sentence and then looking at the globe to see if you understand what the sentence just said.

First: The lines that divide the Earth from North to South are called **meridians** or **lines of longitude.** You can remember that name because they are all the same length. Longitude = long!

- These lines intersect both the North and South Poles and are evenly spaced around the globe.

- They make the globe look as though it had sections as an orange has sections.

- They are farthest apart at the equator, about 80.5 miles, and closer at the poles—as close as you get.

- They run north and south, but you measure the distance between them from east to west. This may sound odd, but it helps to see the same principle on a ruler. Put a ruler in front of you horizontally. The ruler is set to measure left to right. Now imagine that the little lines on the ruler are only parts of longer lines that run vertically. You can see how vertical lines measure horizontal position and distance.

- They measure from the 0° line, which runs through Greenwich, England. The lines divide the Earth into 360 degrees, but they read only up to 180 degrees. This is because the readings go up from 0° in both directions, east and west. The farthest that you can travel away from a place is halfway around the globe. After that you are starting to head back to where you started. In the same way, the farthest you can travel from 0° on the Earth is 180 degrees. Then you start to return on the measurements of the other direction. For instance, if you travel to the Pacific Ocean and get to 179° west, the next marker is going to be 180° west, which is also 180° east. From there you have two choices: go to 179° east or back to 179° west. You can only come back when you get that far away!

Second: The lines that divide the Earth from west to east are called **parallels** or **lines of latitude.** You can remember that name because they are side-to-side, or lateral, lines. Latitude = lateral.

- These lines do not connect but are evenly spaced and always the same distance apart (about 80.5 miles); they are parallel.

- They make the globe look as though it is divided as an onion when it is cut into slices.

- They are not all the same distance around. The ones near the equator are longer, close to 29,000 miles, and the ones near the poles are shorter.

- They run east and west but measure distance and position north and south. Try using the ruler again. This time place the ruler vertically, to measure up and down, and then imagine the little lines on the ruler as parts of longer lateral lines.

- The lines of latitude begin measuring from the 0° line, which is the equator, or the line around the Earth whose points are each equally distant from the North and South Poles.

- Because North and South are places and not just directions, when you start at the 0° marker, the equator, and travel either north or south, you can get only as far away as the North or South Pole before you start coming back. That is only one-quarter of the way around the globe, or 90 degrees. Lines of latitude measure distance north or south from 0° to 90°.

Remember that the Earth spins on an axis, an imaginary rod that goes through the Earth, and that the two points where the axis pierce the surface of the Earth are called North and South. North and South are actual points as well as directions. If you are standing on the North Pole, you can't be any more North, and every other place is South.

Using the Grid
Parallel and Curved Lines

The grid that is on any map is always going to be accurate. The lines of latitude and the lines of longitude stay in the same place no matter how small or large an area the map covers. This sameness is the beauty of the global grid system. It is used in all countries and has become one of the few universal languages that people of the world use to communicate. We make compromises, though, when we use it.

If a map maker were making a map of the world and drew in all the degree and minute lines, there would be no room for anything else. The map would look like a window screen. The second measurement would be impossible to use on a large scale.

Instead of covering a map with lines, a map maker will include as many lines as are necessary to help locate sites accurately. It is up to the map user to do the rest and that requires the skill of dividing by eye.

Try this:

1. Draw a line on a piece of paper and try to divide it into halves. Then with a ruler measure how accurate you were.

2. Try dividing other lines into thirds and quarters and fifths and so on, measuring for accuracy afterward. When you get to smaller divisions, remember that it is easier to divide divisions than to divide all at once. In other words, a twelfth is a third of a half of a half. With a little practice you will be able to train your eye to be fairly accurate, just as carpenters, masons, surveyors, and many other tradespeople have trained their eyes to measure accurately.

3. Look at the grid below. Which hemisphere is it in? If the numbers increase as you go south, then it is in the Southern Hemisphere. If the numbers increase if you go east, the grid is in the Eastern Hemisphere. In reality, this grid would be found over the southwestern tip of Australia.

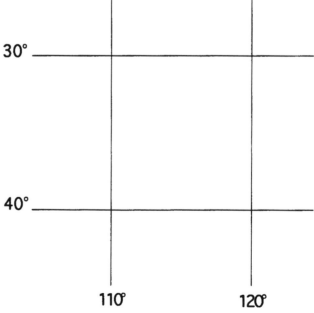

Gumshoe Geography © 1996 Zephyr Press, Tucson, AZ

If we reverse the numbers on the latitude and the longitude scales, where are we? We must be in the Northern Hemisphere and in the Western Hemisphere because the numbers increase as we go north and west. In reality, this grid would be found over Arizona and southern California.

Any site on the globe has two numbers and a direction with each number. This set of numbers and directions is called a set of *coordinates*. The steps taken to find the site follow:

1. Find the box that the site is in by finding the parallels that it is between and the meridians that it is between. These will define a box where the site is located.

2. Do the division that the numbers require. To find sites 1, 2, and 3 on this grid, you will need to divide the sides of the box into tenths.

3. From the correct point on the parallel and the correct point on the meridian, draw perpendicular lines until they cross. You have located the site.

Let's see how this process works using real coordinates. In the grid below, Numbers 1, 2, and 3 represent three cities in the western United States. Site 1 is Phoenix, Arizona. It is located at approximately 33°N, 112°W. You can find it by dividing one of the meridians into tenths by eye. Figure where 112° would be, and trace a line across the grid parallel with the lines of latitude.

Do the same division on the parallels. Find where 33° would be and trace the line up or down until it crosses the 112° line. You are now at our site, Phoenix, Arizona. Can you feel the sun?

Site 2 is Carson City, Nevada. See if you can figure the coordinates. Site 3 is Los Angeles. Figure the coordinates for this site, also. You'll find the answers further on.

The three cities on this grid are found at those locations no matter what map they appear on. That's the beauty of the global grid system. (Oh yes, the coordinates are about 39°N 120°W for Carson City and 34°N 118°W for Los Angeles.)

The map maker might put lines that are in 10° intervals, 15° intervals, or any other interval that fits the map. You will have to read the numbers and decide how you are going to divide the spaces

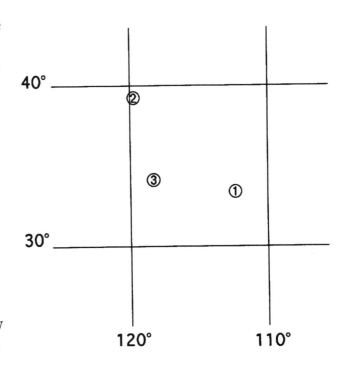

between the lines. The intervals between the parallels are not always the same as the intervals between the meridians, so you have to be careful.

Now, just when you think you're getting it, there's another twist, one more thing to deal with. Our little planet isn't flat, it's spherical. That's great for being part of the solar system, but it makes it nearly impossible to make an accurate flat map of the Earth. People have invented ways to display the globe, called projections, that have come close and are quite usable. We have to remember that the larger the area the map covers, the more out of scale parts of it are apt to be.

Also, there are curved lines on maps. It was easy to divide a line and then follow a course across the grid when all of the lines were straight. That is rarely the case on real maps. The closer you get to the poles or the larger the area that you are studying, the more the lines are going to curve on the flat map. Just remember that no matter how much the map seems to change, you will always be able to find any place at the same coordinates on any map.

One way to deal with the variables is to locate the approximate site on a world map by rounding the coordinates and then turn to a more specific map that covers less area to get closer to the exact location. If local maps aren't available, you will have to do your plotting on the big map using the curved lines. The line that you trace will have to average the curve of the two lines that it's between. Its really not hard; it just takes practice.

Use the blank grids on the pages that follow to practice. Assign the lines numbers in intervals that you select and then practice finding places by yourself, with a partner, with a group, or with your class. Get your teacher involved. (Remind him or her that it's never too late to learn something!)

Happy Plotting!

Parallel Lines

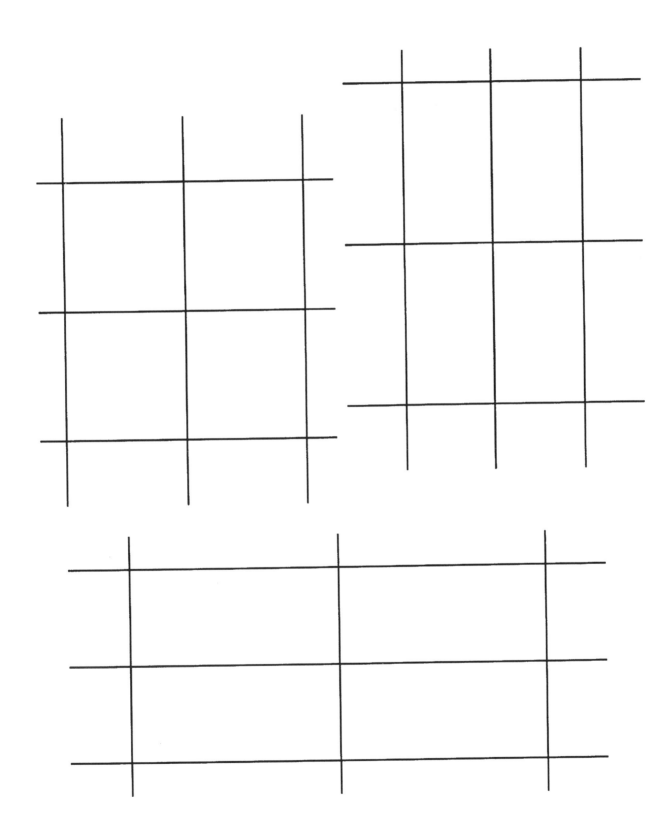

Curved Lines

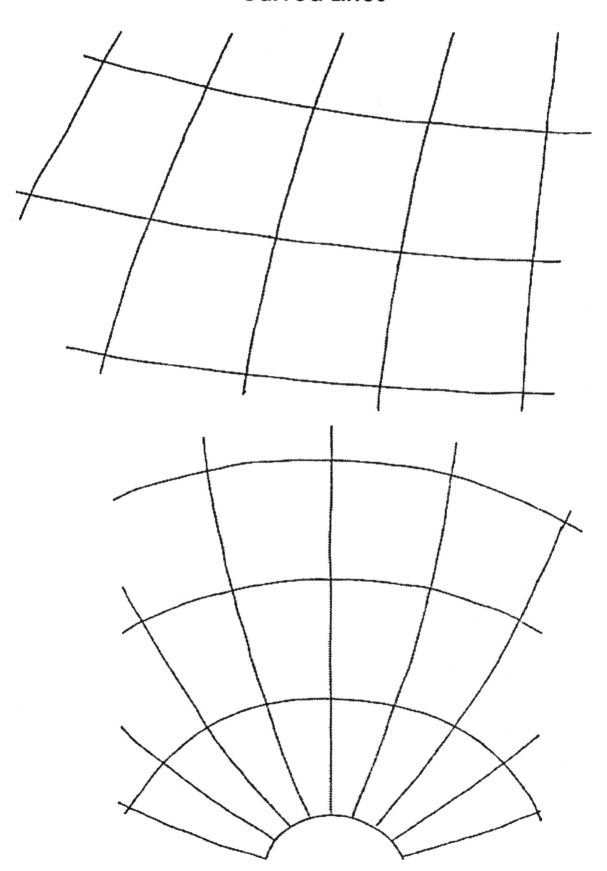

Curved Lines

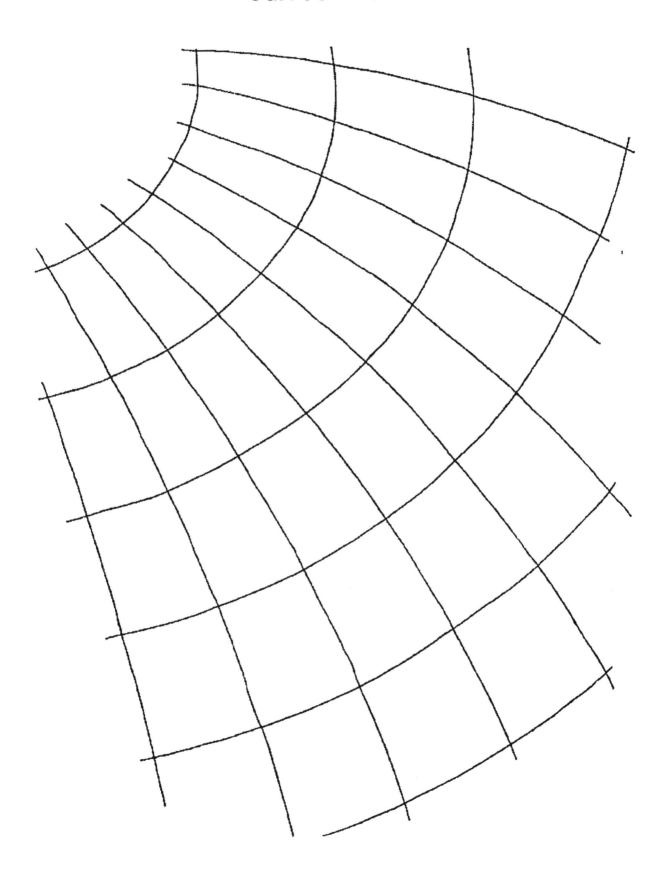

Map Scale and Distance

Measurement is nothing more than dividing any distance, weight, or volume into equal, manageable units and giving the units a name. The name could be miles, gallons, kilograms, or pints; it could be harpals, snippets, plickens, or rulps.

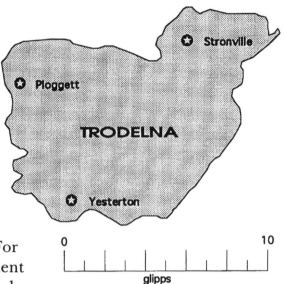

For your map reading you will be using mostly miles and kilometers. They are similar—100 kilometers equals about 62 miles—but what's in a name, anyway? For the sake of practice, we'll use a measurement called a glipp. Cut a strip of paper a few inches long, then let's measure!

Place your strip of paper on the map of Trodelna, with one corner on the star of the town Ploggett. Run the edge of the paper through the town of Yesterton. Make a mark on the strip where it crosses the star at Yesterton. Put the first corner of the strip on the 0 line of the scale and read how far it is between the two towns. It should be about 6.2 glipps (divide each glipp by eye into tenths). Now measure the other two distances. You should get about 8.5 glipps from Ploggett to Stronville and about 9.6 glipps from Stronville to Yesterton.

If you would like to find the distance from Yesterton in Trodelna to Niles in Billiad, you need to measure the opposite way. Place the strip of paper against the scale and copy it onto the strip. Place the beginning of the scale on the star at Yesterton. Aim the edge of the strip toward Niles and make a small mark on the map at 10 glipps. Place the beginning point of the scale copy on the point mark and aim the strip again. Repeat the procedure until the edge of the strip crosses the star at Niles. The last measurement should be from the point mark to where the scale crosses Niles. Add it up and you have the distance!

Under some circumstances, you will have to create your own scale on a blank map. Simply find two places that you know the distance between and use that measure as your standard. Let's say that the distance between Millury and Trundett is 150 sids. Make a scale using that distance, divide it into intervals if you wish, and then use that scale to measure any other distances on the same map. Try it on this map. Compare your results with your partner's or other classmates'. Good luck and happy measuring!

Pronunciation

Have you ever spoken with someone from another country who is just learning your language? Was it hard to understand the person who was trying to pronounce your words? A very big part of communication and relating to other people is being able to pronounce words that express our more complex thoughts, and a very big part of communicating in geographical terms is being able to pronounce geographical names.

It's really not as hard as it might seem at first. Once you get into the swing of it, you'll find that within regions of the world there are place names that have certain similarities. You will get so good at recognizing the similarities that you will be able to find a world region just from hearing a place name.

Some atlases have pronunciation keys. If yours doesn't, you can always use a dictionary. Following are some examples; try them out. They go from easy to difficult. Try also to guess the world region of each name. You'll probably be surprised at how well you do! When you have learned how to say each name, practice it a few times, then practice the entire list for accuracy and speed. You will find that it will be easier for you to remember the location of the place if you can say the name.

Let's try some easy ones first. Here are some names from an English-speaking country:

Belfast	Birmingham	Cardiff
Cork	Dublin	Dundee
Glasgow	Inverness	Liverpool
London	Nottingham	Plymouth

You shouldn't have much trouble with those! Now try some from some northern countries:

Élborg, Denmark	Bergen, Norway	Copenhagen, Denmark
Gîteborg, Sweden	Helsinki, Finland	Murmansk, Russia
Namsos, Norway	Oslo, Norway	Reykjavik, Iceland
Stockholm, Sweden	Tromso, Norway	Vaasa, Finland

Those may be more of a challenge, but you can do it with a little practice. Next are some names from other parts of the world. See how you do.

Following are a few Middle Eastern names. Feel the hot sun and the dry wind when you read them!

Aleppo, Syria	Amman, Jordan	Tripoli, Lebanon
Beirut, Lebanon	Damascus, Syria	Ghazzah, Israel
Irbid, Jordan	Jerusalem, Israel	Baghdad, Iraq
Mosul, Iraq		

Following are some names from some of the Slavic countries in Eastern Europe. They are a real workout!

Bratislava, Slovakia	Brno, Czech Republic	Budapest, Hungary
Prague, Czech Republic	Gyîr, Hungary	Kosice, Slovakia
Krakow, Poland	Maribor, Slovania	Plzen, Czech Republic

Following are a few names from the Far East. Try saying them slowly, and then as you get used to them, speed it up. It's really quite fun and easy.

Alma-Ata, Kazakhstan	Dusanbe, Tajikistan	Hotan, China
Karaganda, Kazakhstan	Kashi, China	Taipei, Taiwan
Bangkok, Thailand	Chiang Rai, Thailand	Hanoi, Vietnam
Dhaka, Bangladesh	Mergui, Burma	Phnom Penh, Cambodia
Rangoon, Burma	Sittwe, Burma	ViangChán, Laos
Changchun, China	Chengdu, China	Dalian, China
Hai Phong, Vietnam	Hakodate, Japan	Kagoshima, Japan
Kobe, Japan	Wuhan, China	Pusan, Korea
Samarkand, Uzbekistan	Semipalatinsk, Kazakhstan	
Taskent, Uzbekistan	Ho Chi Minh City, Vietnam	
Novokazalinsk, Kazakhstan		

Look through your atlas and find other fun places to pronounce. Remember that you'll learn locations better when you're able to pronounce their names.

   Gumshoe Geography © 1996 Zephyr Press, Tucson, AZ

Reading Topical Maps

The most basic use of a map is to show where a specific place is and how close or far it is from other places. Most maps are either in a physical or a political format. The physical format usually shows what the land is like around the site; the political format usually shows more detail of the other sites that are around the chosen site.

Some maps are used to show other things about the world. These maps are called topical maps, and they usually include only a basic outline of the landforms of the area being studied. They are used to show anything from weather patterns to population patterns to locations in which corn is grown.

Look at the topical maps of Alaska below. You can learn many things about Alaska just by looking at these maps. When you begin to use the information from the four maps together you really get to know the place. For instance, you can see by looking at map 3 that both Nome and Anchorage are in the heath zone of natural vegetation. By looking at maps 2, 3, and 4, you might conclude that it's easier to live in Anchorage than in Nome because it's probably easier to raise food in Anchorage. Try making some conclusions of your own by looking at two or three of the maps. You will soon see the value of topical maps.

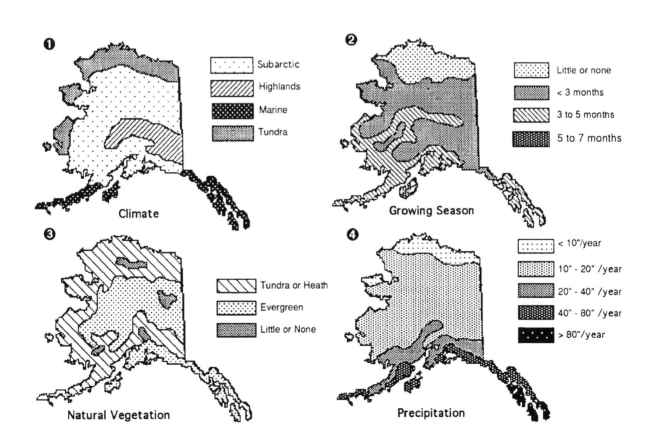

❶
Subarctic
Highlands
Marine
Tundra

Climate

❷
Little or none
< 3 months
3 to 5 months
5 to 7 months

Growing Season

❸
Tundra or Heath
Evergreen
Little or None

Natural Vegetation

❹
< 10"/year
10" - 20" /year
20" - 40" /year
40" - 80" /year
> 80"/year

Precipitation

Training Manual Conclusion

You have passed the training course and have been chosen for service in the elite information-gathering corps of an ultra-secret committee known only as The Committee. You will not know our mission, only that it is peaceful and will support the continuation of life on Earth as we know it. You will not know our location. Your headquarters and starting point for most missions will be where you are.

You will be given assignments and sent on missions to gain information about other cultures and places in the world. The Committee has many needs for the information that you will collect and report. While working for The Committee you may be asked to do the following:

◆ Choose and justify the best mode of travel from one place to the next. You must travel on the surface and not use air travel, because The Committee has decided that air travel is not safe. We also want you to get to know the terrain and the people of the regions on the trip.

◆ Stay healthy for your entire journey, which means choosing appropriate clothing for each leg of the trip and eating nutritionally in all of the places that you visit.

◆ Avoid offending the people with whom you come in contact at each destination, which means learning their customs and culture and, when possible, eating and dressing as they do.

◆ Find the shortest route between all the destinations. The Committee discourages wasted time or resources. You will present a complete report to your supervisor upon your return.

CAUTION
If there are destinations that can be considered danger areas, you must make special notes and provide The Committee with precise information so that we may avoid trouble situations.

Gumshoe Geography © 1996 Zephyr Press, Tucson, AZ

Instructions for Data Report Form

1. Write the numbers of the destinations in the boxes in order. If no starting point is designated, start at your home town. If the order is not specified in the problem, order the destinations to make the trip as efficient as possible.

2. **Coordinates:** Write the coordinates of each destination in these spaces.

3. **Destination:** Write the specific names and the countries or regional names of the places that match the given coordinates.

4. **Mode of Travel:** In these boxes, make note of the ways that you have decided to travel. Air travel is not an option. Walking, human-powered locomotion, and any type of powered surface transportation are options. Combinations are acceptable but they are more difficult to calculate and annotate, so I do not advise them.

5. **Terrain:** Make note in these boxes of the major type or types of landscape that you will encounter on this leg of the journey. What you write here should coordinate with your mode of travel.

6. 🖳 Enter the time that you estimate the journey will take you. This time may be recorded in days for long trips or hours for shorter ones. Be sure to take into account the mode of travel and the terrain. Be prepared to justify your answer.

7. 🧭 Enter overall general direction in these spaces.

8. **Distance:** In these spaces, record the approximate distances that you travel. These distances may be in miles or kilometers or other measure as the problem requires.

9. **Arrival/Departure:** Write the times or dates that correspond to the beginning and end of your trips in these spaces. If you stay for a period in any one place for any reason, record that stay here. A long stay at any destination will make the information about food, clothing, and other cultural and environmental factors more important than if you are just passing through.

10. **Dietary Needs:** In these spaces make note of any special foods or drinks that you think you should or should not consume. Also make note of foods that are required by the cultural setting and those from your normal diet that will not be available at all. In this category as well as the next, it is important to find a balance between fitting in with the local people and staying healthy by maintaining the diet and clothing that your body is used to.

11. **Clothing Needs:** Decide on appropriate clothing for each leg of the trip and enter the choices here. Consider these decisions in the same way that you did the choices of food.

12. **Special Notes:** Anything about the culture or the place that is not mentioned already should be noted here, including information about customs and languages. These notes will make your stay in the various destinations more productive and will help you avoid any unnecessary hassles or delays.

Agent Name: _____   Code Name: _____   Page: _____

Status: _____   Assignment: _____   Date: _____

☐ Coordinates Destination Distance	Mode of Travel Terrain	Arrival Departure	Dietary Needs Clothing Needs Special Notes
☐			
☐			
☐			
☐			

1
ONE COUNTRY

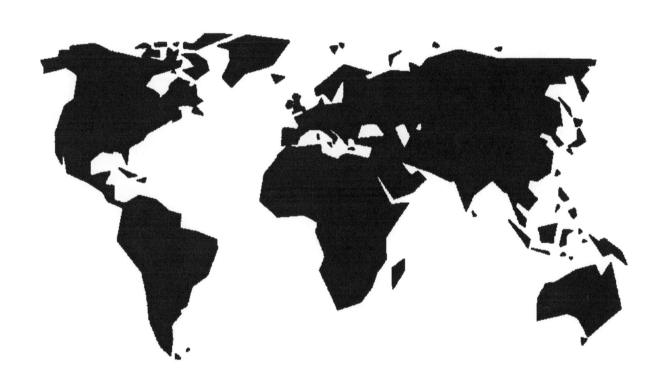

World Map Activity

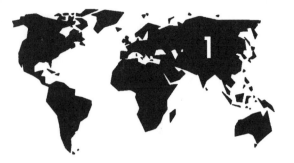

One Country

Your Assignment: The Committee is planning to build a series of microwave transmission towers to make cross-country communication quick and reliable for our members in the host country. Your job is to visit the following locations and determine the risks of placing our towers at the selected sites. Some are rural, some urban. Follow all orders and good luck.

23°02'N	72°37'E	18°58'N	72°50'E
22°32'N	88°22'E	17°23'N	78°28'E
16°56'N	82°13'E	31°35'N	74°18'E
28°36'N	77°12'E	25°36'N	85°07'E
34°05'N	74°49'E	8°29'N	76°55'E

Special Instructions: The Committee has been observing your work and has confidence in your abilities and potential. We would like to give you an opportunity for advancement.

- Please determine the status of the relationship between the host country and each of its neighbors.

- Choose the country that is most likely to be friendly and agree to the expansion of the network on its soil.

Gumshoe Geography © 1996 Zephyr Press, Tucson, AZ

One Country

Your Assignment: The Committee has a client that is in need of certain minerals for a special project. We believe that you will find mineral deposits near the following cities in this very large country. Your job is to visit these locations and document any mineral deposits that may be in the area and what role those deposits play in the local economy. Follow all orders and good luck.

62°05′N 136°18′W	58°46′N 94°10′W
64°04′N 139°25′W	54°56′N 101°53′W
50°40′N 120°20′W	54°05′N 128°38′W
50°03′N 110°40′W	50°23′N 105°23′W
54°01′N 124°01′W	62°27′N 114°21′W

Special Instructions: The Committee has been observing your work and has confidence in your abilities and potential. We would like to give you an opportunity for advancement. The region of the world that you are studying is well known for its large lakes, rugged landscape, and swift, clear rivers.

- Please decide on the best potential location among the sites for a recreational expedition base and justify your choice. If you want to choose a location that is not on our list, please justify your choice.

One Country

Your Assignment: The Committee has been interested in the possibility of establishing operations in the part of the world that you are to visit. There is a distinct mix of cultural heritages in this area and the combination is a source of societal and governmental friction. Your job is to research the causes and current status of the situation and report back to us. Try to compare and contrast the needs and wants of these groups to give us a thorough overview of the situation. Follow all orders and good luck.

48°00'N	66°40'W	49°53'N	74°21'W
48°26'N	71°04'W	53°50'N	79°00'W
49°47'N	92°50'W	48°57'N	54°34'W
51°17'N	80°39'E	57°00'N	61°40'W
46°09'N	60°11'W	48°23'N	89°15'W

Special Instructions: The Committee has been observing your work and has confidence in your abilities and potential. We would like to give you an opportunity for advancement. This area of the globe has many strong rivers and the government has used them to create a great deal of electrical power.

- Please find out about the positive and negative effects of such a large development.

- Find out how it affects the local population as well as the environment.

One Country

Your Assignment: The Committee is compiling a database of historical information about the country to which you are going to be sent. The following cities have something in common that will become plain to you as you travel among them. Your job is to find out why each city has the common factor and report back to us as soon as possible. Follow all orders and good luck.

52°15′N	81°37′W	71°37′N	117°57′W
65°12′N	123°26′W	49°43′N	113°25′W
58°49′N	122°39′W	62°42′N	109°08′W
53°43′N	113°13′W	61°52′N	121°23′W
56°15′N	120°15′W	58°24′N	116°00′W

Special Instructions: The Committee has been observing your work and has confidence in your abilities and potential. We would like to give you an opportunity for advancement. There are many kinds of wildlife in this country that you have explored.

- Find out about several of the larger groups and about their migratory patterns.
- Make a theme map of the region and show the winter and summer homes of the wildlife and their migratory routes.

One Country

Your Assignment: The Committee is finding it more and more difficult to find places to send members on administrative leave. The site needs to be quiet, out of the way, and relaxing. Your job is to look at the following sites and report on their positive and negative characteristics. Compare them in terms of privacy, comfort, accessibility, and quality of resources. Basically, what do these places have to offer? Follow all orders and good luck.

16°51'N	99°55'W	19°14'N	103°43'W
19°51'N	90°32'W	21°05'N	86°46'W
27°59'N	109°56'W	24°10'N	110°18'W
23°13'N	106°25'W	20°58'N	89°37'W
22°13'N	97°51'W	19°12'N	96°08'W

Special Instructions: The Committee has been observing your work and has confidence in your abilities and potential. We would like to give you an opportunity for advancement. People who live near the sea make their living from the sea.

- Compare the ways in which the people in these places make their living.

- Find out about the history of their vocations and how much or how little they have changed over time.

- Also research what the sea has to offer at each of the sites.

Gumshoe Geography © 1996 Zephyr Press, Tucson, AZ

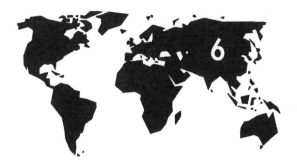

One Country

Your Assignment: The Committee has been looking at ways in which people live in harsh environments. Your job will be to find out about the local fauna and flora in the following places and how the people of each region live there. Please make suggestions as to ways that the quality of life might be improved without stressing the culture. Follow all orders and good luck.

21°53'N 102°18'W	28°30'N 106°00'W
27°37'N 100°44'W	20°40'N 103°20'W
20°58'N 89°37'W	25°41'N 100°19'W
17°00'N 96°30'W	19°19'N 98°14'W
25°33'N 103°26'W	16°45'N 93°07'W
24°02'N 104°40'W	

Special Instructions: The Committee has been observing your work and has confidence in your abilities and potential. We would like to give you an opportunity for advancement. Traveling through this region has many pitfalls and is a hardship if one does not plan adequately and make necessary adjustments.

- After doing research, design a vehicle to travel in this environmental setting that would be versatile and comfortable and would do everything that you would need it to do in your explorations.

- Please make a detailed drawing and describe the vehicle.

One Country

Your Assignment: The Committee has been searching for a source of quality food products to feed our members in remote places and to stockpile for future contingencies. Your job is to visit the following cities and find out about the food produced in the various regions of this country. Report back about any high-quality or unique food products. Follow all orders and good luck.

38°44'S	62°16'W	34°36'S	58°27'W
45°50'S	67°30'W	31°25'S	64°10'W
34°55'S	57°57'W	32°54'S	68°50'W
51°37'S	69°10'W	33°20'S	66°20'W
26°49'S	65°13'W	27°50'S	64°15'W

Special Instructions: The Committee has been observing your work and has confidence in your abilities and potential. We would like to give you an opportunity for advancement.

- When you visit these sites, take a survey of regionally unique vocations and report back about what you find. We would like to match these vocations to certain regionally unique vocations in the United States. Your report will help us do this.

One Country

Your Assignment: The Committee is trying to set new standards for currency in world trade and commerce. We are looking into the use of precious gems as a standard. Your job is to visit the following sites and find out what precious gems are available and how the gem trade is accomplished in this region. Please make recommendations as to how we should proceed. Follow all orders and good luck.

4°36'N	74°05'W	7°08'N	73°09'W
3°53'S	77°04'W	10°25'N	75°32'W
7°54'N	72°31'W	6°15'N	75°35'W
1°08'N	70°03'W	1°13'N	77°17'W
2°27'N	76°36'W	11°15'N	74°13'W

Special Instructions: The Committee has been observing your work and has confidence in your abilities and potential. We would like to give you an opportunity for advancement. This area has a great deal of volcanic and seismic activity.

- Please find out how people live in such an environment and what adjustments they have to make to survive in this region. We would like you to make suggestions as to how the native population might improve its way of life in relation to this activity.

One Country

Your Assignment: The Committee is looking for a stable industrial region where we can have our equipment manufactured. Your job is to visit the following sites and determine the suitability of this region for manufacturing. In your research of the region find out why it is suitable and what natural and locational (cultural, societal, meteorological, recreational, commercial, and so on) resources make it suitable. Follow all orders and good luck.

53°08′N	18°00′E	51°10′N	23°28′E
54°23′N	18°40′E	54°12′N	15°33′E
50°03′N	19°58′E	51°46′N	19°30′E
51°15′N	22°35′E	53°24′N	14°32′E
52°15′N	21°00′E	51°06′N	17°00′E

Special Instructions: The Committee has been observing your work and has confidence in your abilities and potential. We would like to give you an opportunity for advancement.

- Please look into this region's history and find out how stable it has been over the centuries.

- Determine the factors that have played a role in its stability or instability.

- Please make predictions as to its future stability and report back on your findings.

Gumshoe Geography © 1996 Zephyr Press, Tucson, AZ

One Country

Your Assignment: The Committee is interested in how historical land-marks are preserved and how local people feel about preservation in their communities. Your job is to visit the following sites and find out about local monuments and historical buildings in this region, how they are protected, and how the local populace feels about the responsibility. Find out about the history of each site that you find. Follow all orders and good luck.

54°35'N	5°55'W	52°30'N	1°50'W
51°30'N	3°13'W	51°54'N	8°28'W
53°20'N	6°15'W	56°28'N	3°00'W
55°53'N	4°15'W	57°27'N	4°15'W
53°25'N	2°55'W	51°30'N	0°10'W
52°58'N	1°10'W	50°23'N	4°10'W

Special Instructions: The Committee has been observing your work and has confidence in your abilities and potential. We would like to give you an opportunity for advancement. This country is rich in regional heritage.

- Find out about the different national entities that now comprise this country and report on how they are maintaining cultural individuality and how they get along with one another.

- Also report on any unique customs that the various cultural groups continue to follow.

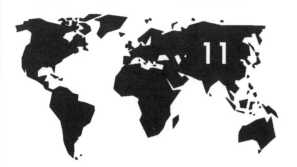

One Country

Your Assignment: The Committee is looking for new customers for its energy holdings and resources. Your job will be to visit the sites listed below and determine how the region satisfies its energy consumption needs. Find out what the local people do for their livings and how energy needs affect their lives. Follow all orders and good luck.

40°49'N	140°45'E	43°46'N	140°22'E
42°45'N	140°43'E	34°24'N	132°27'E
31°36'N	130°33'E	32°47'N	129°56'E
37°55'N	139°03'E	35°37'N	137°14'E
43°03'N	141°21'E	38°15'N	140°53'E
34°21'N	134°03'E	35°40'N	139°46'E

Special Instructions: The Committee has been observing your work and has confidence in your abilities and potential. We would like to give you an opportunity for advancement. This country is a small, crowded country with well-defined family structures.

- Find out what the roles of family members are and how the family structure is changing as a result of global influences and cosmopolitan demands.

- Report on your findings and determine the negative and positive aspects of the changes.

One Country

Your Assignment: The Committee has been watching this country recently. We are interested in how the people cope with losing territory that had provided them with many essential resources. Your job is to visit the following sites and find out how the people are coping with the recent governmental changes. Find out about the availability of the necessities of life and how the people feel about the present state of affairs. Follow all orders and good luck.

69°25′N	86°15′E	48°27′N	135°06′E
62°13′N	129°49′E	56°01′N	92°50′E
59°34′N	150°48′E	55°02′N	82°55′E
53°34′N	142°56′E	55°00′N	73°24′E
54°54′N	69°06′E	56°30′N	84°58′E
66°13′N	169°48′W	43°10′N	131°56′E

Special Instructions: The Committee has been observing your work and has confidence in your abilities and potential. We would like to give you an opportunity for advancement. The people of the region that you are visiting have a rich heritage of resilience in the face of trouble.

- Please find out about and report back on their ability to withstand conquest from the outside, especially in the present century.

One Country

Your Assignment: The Committee is seeking to determine the feasibility of building a transportation/communication corridor across the isolated western region of the country that you are to visit. Your job is to visit the following sites and decide whether it would be a workable project or not. Please report on the natural environment, flora and fauna, at these sites. List and prioritize the factors that brought you to your decision. Follow all orders and good luck.

64°34′N	40°32′E	46°21′N	48°03′E
55°55′N	37°57′E	55°45′N	49°08′E
61°16′N	46°35′E	55°45′N	37°35′E
68°58′N	33°05′E	58°00′N	56°15′E
51°34′N	46°02′E	59°55′N	30°15′E
61°40′N	50°46′E	48°44′N	44°25′E

Special Instructions: The Committee has been observing your work and has confidence in your abilities and potential. We would like to give you an opportunity for advancement. The sites that you visited are mostly isolated by distance, climate, and natural environment.

- Please report on how the people of these communities stay in contact with the outside world and with one another.

- Make suggestions as to how they could improve their communication and their quality of life.

    Gumshoe Geography © 1996 Zephyr Press, Tucson, AZ

One Country

Your Assignment: The Committee is producing a model of a country that is split in half, and two different governments are set up in the opposing halves. Your job is to visit the following sites and report on the two governments in the region and how they have affected a once-united people. Report back on the governments' philosophies and determine which government would be the most receptive to our establishment of bases in their country. Follow all orders and good luck.

41°46′N	129°49′E	39°50′N	127°38′E
42°24′N	128°10′E	37°28′N	126°38′E
40°58′N	126°36′E	35°09′N	126°55′E
38°25′N	127°17′E	35°06′N	129°03′E
37°34′N	127°00′E	39°10′N	127°26′E

Special Instructions: The Committee has been observing your work and has confidence in your abilities and potential. We would like to give you an opportunity for advancement. This region is governed by two opposing governments. We would like for you to find out how the people are getting along in each site.

- Find out how they make a living, how their standards of living compare, and how they are viewed by the world community.

- Decide for yourself which is the best way of life for the people and be sure to justify your decision.

15

One Country

Your Assignment: The Committee has been building a catalogue of native dwellings that are constructed with the natural environment in mind. Your job is to visit the sites listed below and find out about the way the local people construct the shelters that they inhabit. For how long have people been building the traditional dwellings? How did the dwellings come to be as they are today? Follow all orders and good luck.

37°01′N	35°18′E	40°46′N	30°24′E
39°56′N	32°52′E	37°55′N	40°14′E
37°05′N	37°22′E	41°01′N	28°58′E
38°25′N	27°09′E	38°21′N	38°19′E
40°59′N	39°43′E	38°28′N	43°20′E

Special Instructions: The Committee has been observing your work and has confidence in your abilities and potential. We would like to give you an opportunity for advancement. This region is between two continents and two distinct cultures. We are looking at it as a potential site for cross-cultural negotiations.

- Please find out about the culture of the area and determine what parts of the culture are distinctly European and what parts are distinctly Asian. Be ready to defend your answers.

 Gumshoe Geography © 1996 Zephyr Press, Tucson, AZ

One Country

Your Assignment: The Committee is looking for a source for certain strategic raw materials and we believe that they can be found in the region to which you are being sent. Your job is to visit the sites that are listed below and find out about the resources that might be mined or harvested in this island chain. Please report about the availability of those resources compared to their availability in the rest of the world. Please also report about how the gathering of these materials is affecting the natural rain forest environment. Follow all orders and good luck.

3°43'S	128°12'E	1°17'S	116°50'E
6°10'S	106°46'E	1°29'N	124°51'E
3°35'N	98°40'E	0°57'S	100°21'E
0°02'S	109°20'E	7°15'S	112°45'E

Special Instructions: The Committee has been observing your work and has confidence in your abilities and potential. We would like to give you an opportunity for advancement. In this island country there are islands that are heavily populated and islands that are nearly uninhabited.

- Please find out why this disparity exists and what the factors are that make the difference.

- It is reported that the government is trying to relocate some of the people on the crowded islands. Find out why.

- Please make suggestions as to how the situation could be improved. Justify your opinions.

One Country

Your Assignment: The Committee is expanding its list of possible sites for strategic mineral mining. Your job is to visit this country and find out about any sources of minerals or raw materials that we might be able to use in our operations. Report back as to how plentiful they are and where in the region they are mined or harvested. Follow all orders and good luck.

3°18'S	17°20'E	2°30'S	28°52'E
4°18'S	15°18'E	0°25'N	25°12'E
2°09'N	21°31'E	11°40'N	27°30'E
5°49'S	13°27'E	0°04'N	18°16'E

Special Instructions: The Committee has been observing your work and has confidence in your abilities and potential. We would like to give you an opportunity for advancement. This region is the source or the setting for a great wealth of folklore and fable.

- Please find some of the stories about the "jungle" and report back about the truth that these stories address.

- Find out about the people and their customs and some of their history.

- We would like to know how the fables and folklore developed. Please give us your educated opinion after doing the appropriate research.

     Gumshoe Geography © 1996 Zephyr Press, Tucson, AZ

One Country

Your Assignment: The Committee would like to cultivate influence in the country that you are being sent to visit because of its rich deposits of highly valuable minerals. Your job is to visit the following sites and determine where the minerals are, how they are mined, and how we might extend our influence in this region. Follow all orders and good luck.

29°12'S	26°07'E	35°55'S	18°22'E
29°55'S	30°56'E	33°00'S	27°55'E
28°43'S	24°46'E	23°54'S	29°25'E
33°58'S	25°40'E	25°45'S	28°10'E
28°47'S	32°06'E	28°25'S	21°15'E

Special Instructions: The Committee has been observing your work and has confidence in your abilities and potential. We would like to give you an opportunity for advancement. This region has been the site of racial tension between native blacks and white settlers for many years. We would like to know how such a situation evolves.

- Please research the history of white settlement in this region and trace the roots and growth of the conflict.

- Also assess the present situation. We would like you to produce a time line of the major events in this region's history, stressing the causes and effects.

One Country

Your Assignment: The Committee is gathering evidence of poor forest practices for a World Court case in which we are involved. Your job is to visit the following sites and find out about the forest practices in the region and assess whether or not they are conservative or wasteful practices. Find out what the effects on the local environment have been and what these practices tell us about forest practices in the world in general. Follow all orders and good luck.

24°95'S	44°04'E	19°00'S	46°40'E
12°17'S	49°17'E	19°51'S	47°01'E
21°28'S	47°05'E	15°17'S	46°43'E
18°10'S	49°24'E	23°21'S	43°39'E

Special Instructions: The Committee has been observing your work and has confidence in your abilities and potential. We would like to give you an opportunity for advancement. This area of the world is relatively isolated from a large continent and therefore, over the centuries, the species here have developed in unique ways.

- Please report on several of the unusual species and tell how they are adapted to their own niches.

- Discuss the food web in this region and the prey-predator relationships.

- Also find out if any of these species are endangered and how they are being affected by human encroachment in this region.

Gumshoe Geography © 1996 Zephyr Press, Tucson, AZ

One Country

Your Assignment: The Committee is investigating the region to which you are being sent because for centuries it has been one of the centers for music and art and a center for exploration during the centuries following the Middle Ages. Your job is to visit the following sites and find out about the heritage of the region. Try to determine what the factors were that produced so much creative energy in centuries past. Please profile some of the key people in these movements. Follow all orders and good luck.

43°38′N	13°30′E	41°08′N	16°51′E
46°31′N	11°22′E	37°30′N	15°06′E
44°25′N	8°57′E	45°28′N	9°12′E
40°50′N	14°15′E	38°07′N	13°22′E
42°45′N	12°29′E	40°28′N	17°14′E
45°40′N	13°46′E	45°27′N	12°21′E

Special Instructions: The Committee has been observing your work and has confidence in your abilities and potential. The region that you are visiting is comprised of many small, distinct regions that have their own cultures and ways of life.

- For each of the sites that you visit, find out what makes the region around the site distinct and how the people live their lives.

- Please chart the differences among the regions according to the livelihoods of the people, their clothing and shelter, what they produce, how they celebrate holidays, the evidences of their histories, and any other special qualities you might encounter.

- Create a thematic map showing some of the unique features of the separate subcultures.

One Country

Your Assignment: The Committee is exploring the possibility of establishing a laboratory in this region to study agricultural methodology. Your job is to visit the following sites and catalogue the agricultural products of the region around each site. Determine how important each region is to the local economy and make special note of any unique farm practices that you might find. Follow all orders and good luck.

44°50′N	0°34′W	48°42′N	4°29′W
48°00′N	0°12′E	45°45′N	4°51′E
43°18′N	5°24′E	47°13′N	1°33′W
43°42′N	7°15′E	48°52′N	2°20′E
42°41′N	2°53′E	49°15′N	4°02′E
49°26′N	1°05′E	48°35′N	7°45′E

Special Instructions: The Committee has been observing your work and has confidence in your abilities and potential. We would like to give you an opportunity for advancement. This country has produced some of the world's most famous and infamous national leaders.

- Please choose the ten that you believe to be the most influential in this country's history and enter a short biographical sketch of each on a time line.

- Use the information that you find to look for trends in leadership style and focus. Please report back on your findings.

Gumshoe Geography © 1996 Zephyr Press, Tucson, AZ

World Map Activity

One Country

Your Assignment: The Committee is evaluating sites in the Southern Hemisphere for placement of retiring members. The site that you are to visit is a unique site in many ways. It is quite large, it has an active fault running the length of it, and it features many climates within its boundaries. Your job is to visit the sites listed below, research the region, and make a report on any facet of the region that catches your interest. We have many members on site, so please check in before you do your research so that we can avoid duplication of research. Follow all orders and good luck.

36°52′S	174°45′E	43°32′S	172°37′E
45°53′S	170°30′E	38°39′S	178°01′E
37°47′S	175°17′E	46°25′S	168°21′E
41°16′S	173°15′E	38°09′S	176°15′E
44°24′S	171°15′E	41°17′S	174°46′E

Special Instructions: The Committee has been observing your work and has confidence in your abilities and potential. We would like to give you an opportunity for advancement. We have evidence that the educational system at this site is exceptional and that it has experienced many successes.

- Please report on any exemplary learning programs or systems that come out of this region and be prepared to explain why they are so successful.

World Map Activity

One Country

Your Assignment: The Committee is concerned that the stock of domesticated animals in the industrialized nations is becoming too homogeneous. We are looking for new breeds to strengthen the stock. Your job is to visit the following sites and take a survey of the domesticated animals in this region. Please report about how they are suited to their respective environments and what they are used for. Follow all orders and good luck.

23°39'S	70°24'W	18°29'S	70°20'W
36°50'S	73°03'W	41°28'S	72°57'W
51°44'S	72°31'W	53°09'S	70°55'W
34°35'S	71°00'W	33°27'S	70°40'W
39°48'S	73°14'W	33°02'S	71°38'W

Special Instructions: The Committee has been observing your work and has confidence in your abilities and potential. We would like to give you an opportunity for advancement. The people of this region have very distinct musical traditions. We are interested in the instruments that they use.

- Please report on the instruments, the music, and the traditions involved in the making of music in this region.

- If possible, please obtain or construct one of the instruments and learn how to play a simple melody on it.

One Country

Your Assignment: The Committee is looking for new sources for certain strategic minerals. This project is so secret that we can't tell you what minerals we are looking for. Your job is to visit the following sites and report back on any mining and refining activities that are taking place near there. It will become apparent to you which mineral we are looking for. Please also compare this region to other world regions that produce the same mineral. Follow all orders and good luck.

3°10'N	113°02'E	5°25'N	100°20'E
5°59'N	116°04'E	3°10'N	101°42'E
1°33'N	110°20'E	5°50'N	118°07'E
2°18'N	111°49'E	1°17'N	103°51'E

Special Instructions: The Committee has been observing your work and has confidence in your abilities and potential. We would like to give you an opportunity for advancement. The world's food supply is dependent on the safe and successful production of a very few types of grain. We in The Committee are looking to influence the source of one or more of them.

- Please research the grain that is produced in this region and report on all facets of its production.

One Country

Your Assignment: The Committee is investigating the factors that make an otherwise forbidding country attractive for exploration and trade. Your job is to research the factors that first made this region attractive and report on them and on major explorers of the region. Also visit the sites listed below and report on regional products and resources that might attract trade or tourist traffic today. Follow all orders and good luck.

39°55′N	116°23′E	43°51′N	125°20′E
30°45′N	104°04′E	45°45′N	126°37′E
25°08′N	102°43′E	29°42′N	91°07′E
31°59′N	118°51′E	31°41′N	121°28′E
43°48′N	87°35′E	34°15′N	108°50′E

Special Instructions: The Committee has been observing your work and has confidence in your abilities and potential. We would like to give you an opportunity for advancement. The region that you are exploring is a vast and relatively backward region by industrialized Western standards.

- Please i   estigate the transportation systems that are in place there and report on how the problems of getting people and goods from place to place are solved. Be prepared to make recommendations for improvement if appropriate.

One Country

Your Assignment: The Committee is thinking about preparation for natural disasters and is researching water shortages and ways of dealing with them. Your job is to visit the following sites and report on how these people deal with water shortage as a fact of everyday life. Please report back on your findings. Follow all orders and good luck.

34°56'S	138°36'E	23°42'S	133°53'E
27°28'S	153°02'E	17°58'S	122°14'E
12°28'S	130°50'E	28°46'S	114°36'E
30°45'S	121°28'E	37°49'S	144°58'E
20°44'S	139°30'E	31°56'S	115°50'E
33°52'S	151°13'E	19°16'S	146°48'E

Special Instructions: The Committee has been observing your work and has confidence in your abilities and potential. We would like to give you an opportunity for advancement. The region that you are visiting has an interesting history of both its native people and colonization.

- Please report on these issues and tell how they have combined to produce the culture that currently exists in this region.

One Country

Your Assignment: The Committee would like to build a resort community for our members and administrators in the region that you are about to visit. Your job is to visit the following sites and determine the nature of the resources in each. Also investigate the heritage of the region and report on how it influences the local culture in these sites. Follow all orders and good luck.

1°27′S	48°29′W	3°43′S	38°30′W
5°47′S	35°13′W	30°04′S	51°11′W
8°03′S	34°54′W	22°54′S	43°15′W
12°59′S	38°31′W	23°57′S	46°20′W
2°31′S	44°16′W		

Special Instructions: The Committee has been observing your work and has confidence in your abilities and potential. We would like to give you an opportunity for advancement. The region that you are going to visit is spread out and most of the sites are isolated from one another.

- Please investigate the ways that the people communicate with one another in this region.

- Find out about all facets of communication, from private communication to the use of mass media. Be prepared to compare it to communication in your country.

Gumshoe Geography © 1996 Zephyr Press, Tucson, AZ

One Country

Your Assignment: The Committee would like to study the effects of certain climates and ways of life on life expectancy in order to begin to form a policy about lifestyle for our members. We want them to live long and healthy lives. Your job is to visit the following sites and research the statistics on life expectancy, birth rate, causes of death, infant mortality rate, and any others that you find to be significant for our study. Be prepared to tell why you think the numbers are what they are. Follow all orders and good luck.

19°55′S	43°56′W	15°47′S	47°55′W
20°27′S	60°50′W	19°01′S	57°39′S
15°35′S	56°05′W	25°25′S	49°15′W
5°50′N	55°10′W	8°46′S	63°45′W
9°58′S	67°48′W	29°41′S	53°48′W
23°32′S	46°37′W	19°45′S	47°55′W

Special Instructions: The Committee has been observing your work and has confidence in your abilities and potential. We would like to give you an opportunity for advancement. This region is vast and very remote. It is also covered by a great network of rivers.

- Please report on how the local people use the rivers to meet essential needs.

- Please also report on any other resources that have a significant influence on the culture.

One Country

Your Assignment: The Committee would like to learn how the cultures of various peoples differ because of environment and opportunity. Your job is to visit the following sites and find out about the factors that influence regional differences. Please report on those factors and construct a chart that displays your findings. Follow all orders and good luck.

52°31′N	13°24′E	50°44′N	7°06′E
53°05′N	8°48′E	51°03′N	13°45′E
50°55′N	13°22′E	52°22′N	9°43′E
51°18′N	12°20′E	48°09′N	11°35′E
49°27′N	11°05′E	54°05′N	12°08′E
48°46′N	9°11′E	50°05′N	8°15′E

Special Instructions: The Committee has been observing your work and has confidence in your abilities and potential. We would like to give you an opportunity for advancement. This region has a very strong nation-alistic culture.

- We would like for you to find out about the environmental and heritage factors that influence the culture.

- Please report on the negative as well as the positive characteristics and manifestations of this culture.

     Gumshoe Geography © 1996 Zephyr Press, Tucson, AZ

2
WORLD REGIONS

World Map Activity

World Regions

Your Assignment: The Committee is preparing to do broad experimentation on the resources associated with glacial terrain. Your job is to visit the following sites and report back on how the natural environment has been and is now being affected by the glacial age. Be prepared to answer questions about the people who live near the glacial region in this area. Follow all orders and good luck.

57°03′N	9°56′E	60°23′N	5°20′E
55°40′N	12°35′E	57°43′N	11°58′E
60°01′N	24°58′E	68°58′N	33°05′E
64°30′N	11°30′E	59°55′N	10°45′E
64°09′N	21°57′W	59°20′N	18°03′E
69°40′N	19°00′E	63°06′N	21°36′E

Special Instructions: The Committee has been observing your work and has confidence in your abilities and potential. We would like to give you an opportunity for advancement. The region that you are scheduled to visit has one of the world's highest standards of living as well as low mortality and crime rates.

- Please report about the factors that you think influence these statistics and suggest how we might adapt them to other sites.

World Regions

Your Assignment: The Committee is working at creating models for a strong governmental structure and we are interested in studying the ancient philosophies of government that have come from the region to which you are being sent. Your job is to visit the following sites and be prepared to discuss the different forms of government that have held power in this region and make recommendations as to which are adaptable to modern culture. Follow all orders and good luck.

37°59′N	23°44′E	37°02′N	22°07′E
39°40′N	19°45′E	40°37′N	20°46′E
41°21′N	21°34′E	42°05′N	19°30′E
42°00′N	21°29′E	40°38′N	22°56′E
41°20′N	19°50′E	41°08′N	24°53′E

Special Instructions: The Committee has been observing your work and has confidence in your abilities and potential. We would like to give you an opportunity for advancement. The language heritage in this region is rich and varied.

- Please study the most influential languages and comment on their roots, their written and spoken forms, and how the people react to standardization.

- Please also report on any regional dialects that certain local groups might be reluctant to give up.

World Regions

Your Assignment: The Committee is planning to establish a transportation system in a very mountainous region of the world. Your job is to visit the following sites and study how people get around. Be prepared to discuss the history of transportation in this region and the factors that have caused it to change. Follow all orders and good luck.

46°31′N	11°22′E	46°10′N	6°10′E
47°16′N	11°24′E	48°18′N	14°18′E
46°05′N	8°20′E	48°09′N	11°35′E
47°48′N	13°02′E	47°08′N	9°30′E
48°12′N	16°22′E	47°20′N	8°35′E

Special Instructions: The Committee has been observing your work and has confidence in your abilities and potential. We would like to give you an opportunity for advancement. Local foods are of great interest to us.

- We would like you to study and report back on how food products are produced in this environment. We are especially interested in dairy products and their use and production.

World Regions

Your Assignment: The Committee has established control of some of the world's most prolific vineyards and we need a source of cork to complete the package. Your job is to visit the following sites and report on the agricultural activity in general and the production of cork specifically. Be prepared to discuss where in the world it is used and what other uses of cork might be. Follow all orders and good luck.

41°23′N	2°11′E	36°32′N	6°18′W
37°36′N	0°59′W	38°34′N	7°54′W
38°43′N	9°08′W	40°24′N	3°41′W
36°43′N	4°25′W	41°09′N	8°37′W
43°19′N	1°59′W	37°23′N	5°59′W
39°50′N	4°00′W	39°28′N	0°22′W
41°35′N	1°00′W		

Special Instructions: The Committee has been observing your work and has confidence in your abilities and potential. We would like to give you an opportunity for advancement. The region that you are going to visit is not very industrialized.

- We would like for you to report on the effects of an agrarian society on the people's health.

- Please present statistics and an explanation that relates the causes to the effects.

World Regions

Your Assignment: The Committee has been working on the problem of the loss of arable land near desert regions. Your job is to visit the following sites and report on the ways that people deal with the encroachment of the desert into their areas. Be prepared to discuss how serious the problem is and how the people might solve it. Follow all orders and good luck.

24°12′N	23°18′E	36°47′N	3°03′E
24°05′N	32°53′E	17°55′N	19°07′E
19°04′N	8°24′E	31°38′N	8°00′W
20°45′N	17°01′W	33°53′N	10°07′E
22°42′N	3°56′W	27°45′N	8°25′W

Special Instructions: The Committee has been observing your work and has confidence in your abilities and potential. We would like to give you an opportunity for advancement. Because of its harsh climate and environment, this region has a distinct culture.

- Please report on the ways that the people trade with one another and with the outside world, and report on the types of currency that they use.

- Try to find out about the history of currency in this region.

Gumshoe Geography © 1996 Zephyr Press, Tucson, AZ

World Regions

Your Assignment: The Committee is planning a training expedition into a mountainous region. Your job is to visit the following sites and find out how the local people deal with the rough topography and the constant threat of volcanoes and mud slides. Be prepared to make recommendations for an optimal route and itinerary through this region. Follow all orders and good luck.

10°30'N	66°56'W	4°56'N	52°20'W
8°08'N	63°33'W	6°48'N	58°10'W
10°40'N	71°37'W	9°45'N	63°11'W
5°50'N	55°10'W	1°55'N	67°04'W
7°46'N	72°14'W	10°11'N	67°45'W

Special Instructions: The Committee has been observing your work and has confidence in your abilities and potential. We would like to give you an opportunity for advancement. This region was the site of one of the most advanced civilizations in the world.

- Please report on the history and the customs of that civilization, and also report on the influences that that civilization has on the present culture in this area.

World Regions

Your Assignment: The Committee is helping a certain island nation petition the World Court to move the international dateline in order to get the nation into a different zone. Your job is to visit the following sites and be prepared to report on the history and development of time zones, the establishment of the international dateline, and the effect that it has on the people who live near it. Follow all orders and good luck.

25°04′S	130°05′W	13°28′N	144°45′E
21°11′S	159°46′W	1°20′N	173°01′E
9°27′S	159°57′E	21°19′N	157°52′W
7°20′N	134°30′E	22°16′S	166°26′E
21°08′S	175°12′W	26°20′N	127°47′E
17°32′S	149°34′W	18°08′S	178°25′E

Special Instructions: The Committee has been observing your work and has confidence in your abilities and potential. We would like to give you an opportunity for advancement. The region that you are to visit is full of a variety of ecosystems and environments because the islands are isolated from one another.

- Please research the variety of life on the islands in this region and then plan a travel itinerary for our scientists so that they might study the most varied yet representative sites in the region.

World Regions

Your Assignment: The Committee is negotiating to gain influence in the affairs and the administration of the last unsettled region in the world. Your job is to visit the following sites and report on the ways in which the various peoples feel about their involvement in the region. Be prepared to discuss the history of the treaties that control the region and to make recommendations as to how we might become involved. Follow all orders and good luck.

90°00'S	0°00'	66°17'S	110°32'E
66°40'S	140°01'E	75°31'S	26°38'W
77°51'S	166°37'E	70°46'S	11°50'E
60°40'S	44°30'W	62°12'S	58°55'W
70°18'S	2°22'W	77°51'N	166°46'E
69°00'S	39°35'E	78°28'S	106°48'E

Special Instructions: The Committee has been observing your work and has confidence in your abilities and potential. We would like to give you an opportunity for advancement. The process of living in the environment of the region that you are visiting is an art and not many people are fit to do it.

- Please report on how people exist in such a region and chart the strategies that they use to get by.

World Regions

Your Assignment: The Committee is in the process of helping our government decide where to build a large-scale petroleum refinery. Your job is to visit the following sites and find out about the natural resources and how plentiful they are. Be prepared to discuss the politics of the world's use of those resources. Find out how friendly with our government each of the countries is and make a recommendation as to the placement of our facility. Follow all orders and good luck.

36°12′N	37°10′E	31°57′N	35°56′E
33°21′N	44°23′E	33°53′N	35°30′E
33°30′N	36°15′E	31°30′N	34°28′E
32°33′N	35°51′E	31°46′N	35°14′E
36°20′N	43°08′E	34°26′N	35°51′E

Special Instructions: The Committee has been observing your work and has confidence in your abilities and potential. We would like to give you an opportunity for advancement. In order for us to understand the people of this region, we will need to be somewhat knowledgeable about their respective religions.

- Please study the major religions of the area and report on the holidays celebrated by each and the reasons the holidays are observed.

Gumshoe Geography © 1996 Zephyr Press, Tucson, AZ

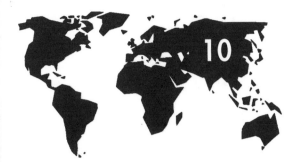

World Regions

Your Assignment: The Committee is doing a study on the effectiveness of natural borders between countries as compared to purely political borders. Your job is to visit the following sites and assess the natural borders that surround the countries you visit. Be prepared to tell whether they have lasted longer than political borders and whether they have been more impregnable to conquest throughout history. Follow all orders and good luck.

33°30′N	36°15′E	58°23′N	26°43′E
41°01′N	28°58′E	50°26′N	30°31′E
53°45′N	27°34′E	55°45′N	37°35′E
43°50′N	18°25′E	44°36′N	33°32′E
41°43′N	44°49′E	52°15′N	21°00′E

Special Instructions: The Committee has been observing your work and has confidence in your abilities and potential. We would like to give you an opportunity for advancement. There are many microcultures in this region.

- We would like for you to identify four and study them.
- Please report on how each small society is suited to the region in which it is located.
- Explain the society's family structure and any other statistics that you think might be important.

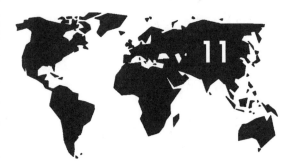

World Regions

Your Assignment: The Committee recognizes that the present time is a rare opportunity to study the process of making a new government work in a country that has gained freedom from a central power. Your job is to visit the following sites and research the types of governments that are being formed there. Be prepared to relate how the people of those countries like their new government and what the new governments are doing for the people. Follow all orders and good luck.

48°09′N	17°07′E	49°12′N	16°37′E
47°30′N	19°05′E	47°41′N	17°38′E
48°43′N	21°15′E	50°03′N	19°58′E
46°33′N	15°39′E	49°45′N	13°24′E
50°05′N	14°26′E		

Special Instructions: The Committee has been observing your work and has confidence in your abilities and potential. We would like to give you an opportunity for advancement. Much of what has gone on in these countries in the way of local culture has been suppressed or kept from the outside world for a long time.

- Please identify some interesting facets of some of the cultural groups in this region and report on them.

Gumshoe Geography © 1996 Zephyr Press, Tucson, AZ

World Regions

Your Assignment: The Committee is always interested in new uses for inhospitable lands. Your job is to visit the following sites and report on what those countries are doing in the region that they all have in common. Be prepared to discuss the use of any resources found in this area and also be prepared to make an assessment of the overall value of this region. Follow all orders and good luck.

82°30′N	62°00′W	69°39′N	162°20′E
71°17′N	156°47′W	69°20′N	53°35′W
78°50′N	103°30′W	71°00′N	8°30′W
78°13′N	15°38′E	76°15′N	119°30′W
68°58′N	33°05′E	69°42′N	170°17′E
77°35′N	69°40′W	71°36′N	128°48′E

Special Instructions: The Committee has been observing your work and has confidence in your abilities and potential. We would like to give you an opportunity for advancement. We have been studying the effects on the body of depriving it of certain essentials. Sunlight is one of those essentials.

- We would like you to find out how the people of this region are affected by the very little sunlight in the winter.

- Please find out how they deal with this situation and what strategies they use to survive.

World Regions

Your Assignment: The Committee has for many years been looking for a way to break down trade barriers among nations and further the establishment of a worldwide open-trade system. Your job is to visit the following sites and find out how the open-trade zone in this region works and if it is a model that we should follow. Be prepared to discuss the possibility of replicating this model in other parts of the world and possibly as a whole world system. Follow all orders and good luck.

57°10'N	2°01'W	52°22'N	4°54'E
50°38'N	5°34'E	60°23'N	5°20'E
44°50'N	0°34'W	48°42'N	4°29'W
55°28'N	8°27'E	53°16'N	9°03'W
43°32'N	5°40'W	50°23'N	4°10'W

Special Instructions: The Committee has been observing your work and has confidence in your abilities and potential. We would like to give you an opportunity for advancement.

- Since much of Western culture was founded here, please find out about the schools in this region.

- Find out how they stack up against schools in the rest of the world.

- Find out what makes a good school and which schools are performing the best and upholding the highest standards.

- State your opinion about the factors that make a school a good one.

Gumshoe Geography © 1996 Zephyr Press, Tucson, AZ

World Regions

Your Assignment: The Committee is helping one of our constituent governments set up a crop bank to preserve hearty species for future generations. Your job is to visit the following sites and catalogue the export crops that are prominent in the region that you are visiting. Be prepared to make recommendations as to which crops are worth banking and to report on how the present crops are suited to certain regional characteristics. Follow all orders and good luck.

41°23′N	2°11′E	37°36′N	0°59′W
37°30′N	15°06′E	44°25′N	8°57′E
35°20′N	25°08′E	39°27′N	2°35′E
38°15′N	21°44′E	42°41′N	2°53′E
40°41′N	14°47′E	40°28′N	17°14′E
43°07′N	5°56′E	35°45′N	14°31′E

Special Instructions: The Committee has been observing your work and has confidence in your abilities and potential. We would like to give you an opportunity for advancement. We are interested in how the change of seasons affects the way of life in this region. It might appear that life in the region is the same all year round.

- Find evidence of cultural patterns that follow the change of seasons.

World Regions

Your Assignment: The Committee has noticed in rural areas of the world a trend away from the use of domesticated animals for labor and toward the use of machinery. We would like to document and possibly preserve any hardy breeds of animal that you can find. Your job is to visit the following sites and find out what animals the people use for labor and for food products. Be prepared to discuss how the animals are used and decide whether the breeds are unique and warrant preservation. Follow all orders and good luck.

40°23′N	49°51′E	41°38′N	41°38′E
39°55′N	41°14′E	43°20′N	45°42′E
40°11′N	44°30′E	40°40′N	46°22′E
45°02′N	39°00′E	38°05′N	46°18′E
41°43′N	44°49′E	40°59′N	39°43′E

Special Instructions: The Committee has been observing your work and has confidence in your abilities and potential. We would like to give you an opportunity for advancement. The people of this region are a stalwart people and they live close to the land.

- Please find out about life expectancy for males and females, and find out enough about the people's lifestyle to determine what factors most affect that statistic.

Gumshoe Geography © 1996 Zephyr Press, Tucson, AZ

World Regions

Your Assignment: The Committee is doing an investigative study on the history and the effects of Western influence in non-Western native cultures. Your job is to visit the following sites and report on the original native cultures and how Western culture has shown its influence in the cities and the rural areas near these sites. Be prepared to discuss the history of the growing Western influence in the sites. Follow all orders and good luck.

5°33′N	0°13′W	1°51′N	9°45′E
7°41′N	5°02′W	4°03′N	9°42′E
11°00′N	7°30′E	6°30′N	3°30′E
0°23′N	9°27′E	6°19′N	10°48′W
9°24′N	0°50′W	3°52′N	11°31′E

Special Instructions: The Committee has been observing your work and has confidence in your abilities and potential. We would like to give you an opportunity for advancement. Attitudes about the participation of women in the daily life of a society differ widely throughout the world.

- Please find out about the attitude toward and roles of women in the cultures of these sites.

- Report on any factors that you believe influence those attitudes.

World Regions

Your Assignment: The Committee is very interested in this region's government policies concerning game and habitat preservation. Your job is to visit the following sites and find out how each government feels about its living resources and in what ways the policies work or don't work. Be prepared to list the major animal species that are protected and how protection affects their food web. Follow all orders and good luck.

3°23′S	29°22′E	6°11′S	35°45′E
0°19′N	32°35′E	1°57′S	30°04′E
2°03′N	45°22′E	4°03′S	39°40′E
2°30′S	32°54′E	1°17′S	36°49′E
5°04′S	39°06′E	6°10′S	39°11′E

Special Instructions: The Committee has been observing your work and has confidence in your abilities and potential. We would like to give you an opportunity for advancement. The native groups that inhabit these regions benefit from the protective policies of the governments in that they are left fairly undisturbed.

- Please find out about their ways of life, particularly about in what ways they meet their essential needs of food, clothing, and shelter.

Gumshoe Geography © 1996 Zephyr Press, Tucson, AZ

World Regions

Your Assignment: The Committee has been in the process of designing models for dealing with drought in nonindustrialized countries. Your job is to visit the following sites and let us know how these places deal with the issue of fresh water, both for irrigation and personal use. Be prepared to report on the new and developing uses of technology that are evident in this region. Follow all   and good luck.

24°28′N	54°22′E	12°46′N	45°01′E
14°48′N	42°57′E	29°20′N	47°59′E
25°17′N	51°32′E	23°29′N	58°33′E
21°27′N	39°49′E	24°28′N	39°36′E
24°38′N	46°43′E	28°23′N	36°35′E

Special Instructions: The Committee has been observing your work and has confidence in your abilities and potential. We would like to give you an opportunity for advancement. The people of this region have a nomadic heritage; yet, because of the presence of petroleum reserves in the region, the people are fairly wealthy.

- Please find out how they deal with the wealth and what the people do with it.
- Find some interesting stories that will illustrate your points.

World Regions

Your Assignment: The Committee has been contracted by one of the governments in the Middle East to help ensure their position of power in the region. Your job is to visit the following sites and report on the history of foreign colonization and domination in this region. Be prepared to discuss the continued effects of the colonial period of history on these countries. Follow all orders and good luck.

30°10'N	48°50'E	37°57'N	59°23'E
34°20'N	62°12'E	25°38'N	57°46'E
34°30'N	69°00'E	30°17'N	57°05'E
37°36'N	61°50'E	36°18'N	59°36'E
30°12'N	67°00'E	34°48'N	48°30'E

Special Instructions: The Committee has been observing your work and has confidence in your abilities and potential. We would like to give you an opportunity for advancement. Women in this culture have a very subdued and hidden role.

- Find out as much as you can about that role and about the family structure in the traditional cultural setting.

- Report on how that role is changing and where the changes are taking place.

Gumshoe Geography © 1996 Zephyr Press, Tucson, AZ

World Regions

Your Assignment: The Committee is doing research on population control and distribution. Your job is to visit the following sites and determine the factors that cause certain areas to be densely populated and others to be uninhabited. Be prepared to report on how certain populations adapt to certain climatic and environmental regions. Follow all orders and good luck.

43°15'N	76°57'E	38°35'N	68°48'E
37°07'N	79°55'E	49°50'N	73°10'E
39°29'N	75°58'E	45°50'N	62°10'E
39°40'N	66°58'E	50°28'N	80°13'E
41°20'N	69°18'E	43°48'N	87°35'E

Special Instructions: The Committee has been observing your work and has confidence in your abilities and potential. We would like to give you an opportunity for advancement. Several of your fellow committee members have filed requests to be placed in this region.

- Please determine the personality and physiological characteristics of four committee members who would influence the site in which they are to be placed.

- Make recommendations about that placement. Please justify your choices.

World Regions

Your Assignment: The Committee is preparing a report on current forestry practices and is interested in the region to which you are being sent. Your job is to visit the following sites and report on the products harvested from the forest and their uses. Be prepared to make a judgment as to whether the countries are practicing beneficial or destructive forest management. Please also relate how the yearly monsoon season affects the industry. Follow all orders and good luck.

13°45'N	100°31'E	19°45'N	99°50'E
16°04'N	108°13'E	23°43'N	90°25'E
21°02'N	105°51'E	10°45'N	106°40'E
22°00'N	96°05'E	12°26'N	98°36'E
11°33'N	104°55'E	16°47'N	96°10'E
20°09'N	92°54'E	17°58'N	102°36'E

Special Instructions: The Committee has been observing your work and has confidence in your abilities and potential. We would like to give you an opportunity for advancement. The region that you are to visit has had a long history of foreign domination and colonialism.

■ Please find out which countries controlled these countries and if there is still influence from those powers.

World Regions

Your Assignment: The Committee has been associated with other agencies that are looking for a solution to world hunger through aquaculture. Your job is to visit the following sites and report on the diet of the local people as it relates to marine products. Be prepared to let us know how they use the ocean to feed themselves and whether they are taking care of the resource. Follow all orders and good luck.

43°51′N	125°20′E	30°45′N	104°04′E
28°55′N	121°39′E	20°52′N	106°41′E
42°45′N	140°43′E	31°36′N	130°33′E
34°41′N	135°10′E	14°35′N	121°00′E
35°06′N	129°03′E	36°05′N	120°21′E
25°03′N	121°30′E	30°30′N	114°20′E

Special Instructions: The Committee has been observing your work and has confidence in your abilities and potential. We would like to give you an opportunity for advancement. The people of this region had a highly developed culture and society long before the people in the rest of the world.

- Please study the ancient cultures in these regions and report on the evidence of advanced culture that the people enjoyed.

- Please tell us about the factors that made the situation change and about the history of the changes.

World Regions

Your Assignment: The Committee has influence in certain industrial areas and we would like to know what is happening in other industrial regions of the world. Your job is to visit the following sites and investigate and report about the industrial activity that is taking place in that region. Be prepared to discuss the region's uses of resources and the people's record on pollution and the environment. Follow all orders and good luck.

47°45′N	26°40′E	44°26′N	26°06′E
44°11′N	28°39′E	44°19′N	23°48′E
45°27′N	28°03′E	47°10′N	27°36′E
42°09′N	24°45′E	43°50′N	25°57′E
45°48′N	24°09′E	42°43′N	23°25′E
45°45′N	21°13′E	43°13′N	27°55′E

Special Instructions: The Committee has been observing your work and has confidence in your abilities and potential. We would like to give you an opportunity for advancement. We have clients who are interested in investing in the growing recreational resort market in the world. There are potential resort markets in this region and we would like you to look into it for us.

- Please make a survey of established recreational facilities in the region and note their season of use.

- Please also recommend sites for future investment.

    Gumshoe Geography © 1996 Zephyr Press, Tucson, AZ

World Regions

Your Assignment: The Committee is beginning to catalogue the factors that governments have valued in other people's lands in centuries past. The region that you are to visit has no strong natural borders and is not very rugged terrain. Your job is to visit the following sites and be prepared to discuss the historically valuable and strategic places and areas of this region and also to discuss why they were considered valuable. Follow all orders and good luck.

52°06′N	23°42′E	51°30′N	31°18′E
48°00′N	37°48′E	52°25′N	31°00′E
50°00′N	36°15′E	50°26′N	30°31′E
46°59′N	28°52′E	53°45′N	27°34′E
47°33′N	30°41′E	46°28′N	30°44′E
46°50′N	29°37′E	55°12′N	30°11′E

Special Instructions: The Committee has been observing your work and has confidence in your abilities and potential. We would like to give you an opportunity for advancement. The one great geographical asset of this region is the presence of a network of rivers.

- Please list the major rivers, then find out how the people use the rivers and why they are important to the culture of the region.

World Regions

Your Assignment: The Committee has members in various parts of the world looking for industrialized areas with quality port facilities. Your job is to visit the following sites and, after you have studied the region, to let us know what industries are there and how they are competing in the world marketplace. Be prepared to tell where raw materials come from and how the finished products get to market. Follow all orders and good luck.

56°39′N	23°41′E	63°45′N	23°54′E
55°43′N	21°07′E	56°35′N	21°01′E
59°23′N	28°11′E	58°42′N	24°32′E
56°57′N	24°06′E	59°25′N	24°45′E
58°23′N	26°45′E	54°41′N	25°19′E

Special Instructions: The Committee has been observing your work and has confidence in your abilities and potential. We would like to give you an opportunity for advancement. This area has a population with a strong cultural tie to the ancient people who lived here. It has also been dominated by Germanic and Russian peoples at various times in its history.

- Please report on how those outside cultures affected the culture of the original people and if the influence can still be found within their culture.

Gumshoe Geography © 1996 Zephyr Press, Tucson, AZ

World Regions

Your Assignment: The Committee is helping a client find a new base of operations. We are looking into establishing it at one of the following sites for reasons that will become obvious. Your job is to visit the sites and report back on what type of government is in place in each of the sites. Be prepared to explain your insight about the friendliness of the people in each site toward outside influence. Follow all orders and good luck.

26°00′N	50°29′E	16°00′N	24°00′W
21°30′N	80°00′W	35°00′N	33°00′E
19°00′N	72°25′W	65°00′N	18°00′W
53°00′N	8°00′W	38°00′N	137°00′E
19°00′S	46°00′E	35°50′N	14°30′E
5°00′S	140°00′E	41°00′S	174°00′E
13°00′N	122°00′E	7°40′N	80°50′E

Special Instructions: The Committee has been observing your work and has confidence in your abilities and potential. We would like to give you an opportunity for advancement. Because of the physical nature of these sites, they all have distinctive animal and plant life, different from that of other places on the globe.

- Please find specific examples of unique animal and plant life in these sites and report on several of them.

- Please choose the ones that you think are the most unusual or special and be sure to tell us why you chose them.

World Regions

Your Assignment: The Committee is looking for centralized sites in various regions of the world in which to establish several major tele-communications bases. Your job is to visit the following sites and find out how the countries get along with their neighbors. Tell what all of the countries have in common and be prepared to discuss how that commonality affects their relationships with their neighbors. Follow all orders and good luck.

33°00′N	65°00′E	47°30′N	14°00′E
53°50′N	28°00′E	17°00′S	65°00′W
22°00′S	24°00′E	15°00′N	19°00′E
47°00′N	20°00′E	31°00′N	36°00′E
18°00′N	105°00′E	47°00′N	29°00′E
47°00′N	104°00′E	23°00′S	58°00′W
46°00′N	8°30′E	1°00′N	32°00′E

Special Instructions: The Committee has been observing your work and has confidence in your abilities and potential. We would like to give you an opportunity for advancement. The following sites are likely to show the influence of many cultures.

■ Please find out about the various cultures that influence daily life in the sites that you are visiting and tell us about the most prominent ones.

■ Please let us know also if one or more of the sites have a culture strong enough to stand alone.

Gumshoe Geography © 1996 Zephyr Press, Tucson, AZ

World Regions

Your Assignment: The Committee has developed a model for international relations and would like to find a site to implement it. Your job is to visit the following sites and find out how these countries get along with their neighbors. Be prepared to discuss how they are aligned and how having so many neighbors affects the culture there. Let us know which neighbors are the best neighbors and which ones are not so good in your opinion. Follow all orders and good luck.

15°00′N	30°00′E	35°00′N	105°00′E
39°00′N	35°00′E	46°00′N	2°00′E
51°00′N	9°00′E	6°00′S	35°00′E
6°30′S	13°30′E	60°00′N	100°00′E
9°00′S	53°00′W		

Special Instructions: The Committee has been observing your work and has confidence in your abilities and potential. We would like to give you an opportunity for advancement. The following sites are surrounded by many neighbors.

- Please tell us if the neighbors influence the cultures of their neighbors or if the opposite is true.

- Please let us know what factors play an important role in the direction or the strength of the influence.

3

UNITED STATES REGIONS

World Map Activity

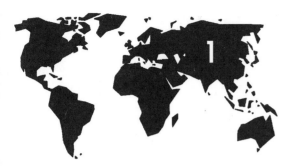

United States Regions

Your Assignment: The Committee is interested in gaining influence in the marine resources industry in the region that you are to visit. The Georges Bank has been one of the richest fishing and shellfish grounds in the world for several hundred years. Your job is to visit the following sites and be prepared to give a full report on the history and the present condition of the marine resources industry there. Follow all orders and good luck.

44°49'N	68°47'W	44°29'N	71°10'W
42°21'N	71°04'W	41°11'N	73°11'W
44°28'N	73°14'W	46°52'N	68°01'W
41°46'N	72°41'W	43°39'N	70°17'W
43°03'N	70°47'W	41°50'N	71°25'W
43°37'N	72°59'W	42°07'N	72°36'W

Special Instructions: The Committee has been observing your work and has confidence in your abilities and potential. We would like to give you an opportunity for advancement. The region that you visited has a rich heritage.

- Please investigate five sites that have historical significance in this region, and record the personal stories of people and events in each place at the time of historical significance.

   Gumshoe Geography © 1996 Zephyr Press, Tucson, AZ

United States Regions

Your Assignment: The Committee has been contracted to do a study of a heavily populated and industrialized region. We have chosen the region to which you are going to be sent. Your job is to visit the following sites and report on the growth of each megalopolis from the early cities that are their centers. Be prepared to explain how each got to be the way it is, what factors keep it this way, and what its future is. Follow all orders and good luck.

42°39′N	73°45′W	39°27′N	74°35′W
39°17′N	76°37′W	42°54′N	78°53′W
38°21′N	81°38′W	39°10′N	75°32′W
40°44′N	74°11′W	38°40′N	76°14′W
40°26′N	80°00′W	37°16′N	79°57′W
41°24′N	75°40′W	40°05′N	80°43′W

Special Instructions: The Committee has been observing your work and has confidence in your abilities and potential. We would like to give you an opportunity for advancement. This region is the first great melting pot for American culture.

- Please give a short report on the different periods of immigration to this region and tell who came when, why they came, where they went, and what they did for a living.

- Report whether the distinct groups exist or to what extent they have been assimilated into mainstream American culture.

United States Regions

Your Assignment: The Committee is assembling information to be published in a booklet of recreational opportunities for our members on leave. Your job is to visit the following sites and investigate what recreational opportunities are available in or near them. Be prepared to describe the sites and tell if they are historical, recreational, or leisure sites, or a combination of these. Follow all orders and good luck.

30°23′N	91°11′W	33°31′N	86°49′W
32°48′N	79°57′W	35°59′N	78°54′W
32°18′N	90°12′W	35°58′N	83°56′W
35°08′N	90°03′W	25°46′N	80°12′W
28°32′N	81°23′W	30°25′N	87°13′W
32°04′N	81°05′W	32°30′N	93°45′W

Special Instructions: The Committee has been observing your work and has confidence in your abilities and potential. We would like to give you an opportunity for advancement. This area has a good portion of low and swampy land.

- Please report on the people who live near the swamps in this region.

- Tell us what they do for a living and also let us know about their family lives.

- Find out about the character of each of the low regions and describe any major variances.

     Gumshoe Geography © 1996 Zephyr Press, Tucson, AZ

4

United States Regions

Your Assignment: The Committee is looking into new areas, especially where there are plentiful raw materials. Your job is to visit the following sites, find what raw materials are available and which ones are the most plentiful, and then be prepared to discuss how the area could be managed if we were to gain influence in the region. Follow all orders and good luck.

37°00′N	86°27′W	39°06′N	84°31′W
39°51′N	89°32′W	45°00′N	87°30′W
38°03′N	84°30′W	42°43′N	87°48′W
42°17′N	89°06′W	43°25′N	83°58′W
46°30′N	84°21′W	48°00′N	88°00′W
39°28′N	87°24′W	41°39′N	83°32′W

Special Instructions: The Committee has been observing your work and has confidence in your abilities and potential. We would like to give you an opportunity for advancement. The economy of this region is heavily dependent on mining and other related industries.

- Please profile three families, real or fictional, who depend on the mines for their living.

- Tell us about their daily lives and also about a typical year.

United States Regions

Your Assignment: The Committee is working with certain environmental groups that are trying to slow forest depletion in this country. Your job is to visit the following sites and find out what the people use for most of their residential building supplies. Be prepared to discuss the historical perspective on building in this region and how economical transportation has changed the way people do things. Follow all orders and good luck.

41°59'N	91°40'W	37°45'N	100°00'W
46°52'N	96°48'W	37°06'N	94°31'W
34°44'N	92°15'W	44°22'N	100°21'W
33°30'N	111°56'W	42°30'N	96°23'W
45°33'N	94°10'W	36°09'N	95°58'W
48°09'N	103°37'W		

Special Instructions: The Committee has been observing your work and has confidence in your abilities and potential. We would like to give you an opportunity for advancement. The people who settled in this area at the end of the nineteenth century faced many hardships, including finding or making appropriate shelter.

- Please report on their solutions to this problem. How were the houses made and how well did they hold up and provide protection from the elements throughout the year?

- Make a drawing of a typical shelter from that time period.

  Gumshoe Geography © 1996 Zephyr Press, Tucson, AZ

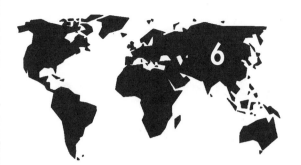

United States Regions

Your Assignment: The Committee is interested in the controversy over the uses and abuses of public land. Your job is to visit the following sites and find out just how much public land there is and what it is used for. Be prepared to say whether you think there are abuses of the land and how things could be done differently. Follow all orders and good luck.

40°01′N 105°17′W	42°51′N 106°19′W
39°05′N 108°33′W	47°30′N 111°17′W
41°19′N 105°35′W	46°25′N 117°01′W
41°44′N 111°50′W	46°52′N 114°01′W
40°14′N 111°39′W	42°34′N 114°28′W

Special Instructions: The Committee has been observing your work and has confidence in your abilities and potential. We would like to give you an opportunity for advancement.

- Please tell us about the history of cattle ranching in this area.
- Choose one era of that industry and write a personal account of a person who worked in the industry.

United States Regions

Your Assignment: The Committee is looking at the process of setting aside public land as national parks in certain developing countries. Your job is to visit the following sites and report on the parks that are on or near the sites. Be prepared to talk about the way in which they were set aside and the reasons for their designation as parks. Tell about the politics involved in founding parks and recommend ways of doing it and problems to watch out for. Follow all orders and good luck.

35°05′N 106°40′W	39°10′N 119°46′W
34°24′N 103°12′W	37°16′N 107°53′W
31°45′N 106°29′W	39°15′N 114°53′W
35°12′N 111°39′W	27°31′N 99°30′W
33°35′N 101°51′W	32°13′N 110°58′W
31°55′N 97°08′W	32°43′N 114°37′W
35°13′N 101°49′W	

Special Instructions: The Committee has been observing your work and has confidence in your abilities and potential. We would like to give you an opportunity for advancement. This region has a long border with a single foreign neighbor.

- Please report on relations with that neighbor and also report on any problems that the two countries might be having.

- Let us know about any significant agreements the two countries have and let us know how those agreements affect the local populations.

 Gumshoe Geography © 1996 Zephyr Press, Tucson, AZ

United States Regions

Your Assignment: The Committee is helping a certain government develop plans and systems for dealing with natural disasters. Your job is to visit the following sites and report on the natural disasters that the region has experienced in the last century. Be prepared to show how the people of the region have coped with the disasters and how they have put their lives back together after them. Speculate as to why people stay in such dangerous places. Follow all orders and good luck.

46°11'N 123°50'W	35°23'N 119°01'W
48°46'N 122°29'W	44°02'N 123°05'W
40°47'N 124°09'W	36°45'N 119°45'W
42°13'N 121°46'W	32°43'N 117°09'W
35°17'N 120°40'W	47°40'N 117°23'W
37°57'N 121°17'W	46°36'N 120°31'W

Special Instructions: The Committee has been observing your work and has confidence in your abilities and potential. We would like to give you an opportunity for advancement. This region of the country has always been viewed as a trend-setting region.

- Please find out about how the people of this region view the education systems in their respective states; find out how they fund it, what they think is important, how they handle higher education, and what new things they are doing for public education.

United States Regions

Your Assignment: The Committee has been working with an eastern government regarding the re-establishment of control in certain parts of the world. Your job is to visit the following sites and report on the natural resources that are available in this region. Be prepared to discuss how plentiful, how accessible, and how valuable they are. Follow all orders and good luck.

61°13'N 149°53'W	71°17'N 156°47'W
59°02'N 158°29'W	53°53'N 166°32'W
64°51'N 147°43'W	66°34'N 145°14'W
55°21'N 131°35'W	64°30'N 165°24'W
59°28'N 185°19'W	61°07'N 146°16'W

Special Instructions: The Committee has been observing your work and has confidence in your abilities and potential. We would like to give you an opportunity for advancement. The region that you are to visit has had a colorful past, and the people who live there reflect that past.

■ Please construct an annotated time line of the history of settlement in this region and discuss the factors that caused population growth at different times in its history.

Gumshoe Geography © 1996 Zephyr Press, Tucson, AZ

United States Regions

Your Assignment: The Committee is looking for an island location for our new top-secret satellite tracking base. Your job is to visit the following sites and make an assessment of the climate and weather patterns there to determine the suitability of each site for our installation. Be prepared to make a recommendation for placement of the facility and to justify your choices. Follow all orders and good luck.

13°28′N	144°45′E	18°21′N	64°56′W
17°45′N	64°40′W	19°44′N	155°05′W
21°19′N	157°52′W	7°20′N	134°30′E
28°13′N	177°22′W	14°16′S	170°42′W
18°28′N	66°07′W	19°18′N	166°36′W

Special Instructions: The Committee has been observing your work and has confidence in your abilities and potential. We would like to give you an opportunity for advancement. Our government has always taken an anti-imperialist stance and yet we have incorporated these sites into our national dominion.

- Please do some research and find out how these sites came to be under our control and how the people of these places feel about having our government as theirs.

- Please also find out if there have been other sites under our control that are now independent.

4
CONTINENTAL

World Map Activity

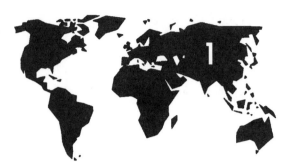

Continental

Your Assignment: The Committee is preparing to retire some of its members and wants to locate them in cities where they will be healthy and out of danger. Your job is to visit each of the following locations and file a comprehensive report on the environmental and cultural data. We will then make the appropriate decisions. Follow all orders and good luck.

9°01'N	38°46'E	4°22'N	18°35'E
19°50'S	34°52'E	12°00'N	8°31'E
15°50'N	33°00'E	0°06'S	34°45'E
31°38'N	8°00'W	4°48'S	11°51'E
16°46'N	2°59'W	24°01'S	21°43'E

Special Instructions: The Committee has been observing your work and has confidence in your abilities and potential. We would like to give you an opportunity for advancement.

- Please research common ailments and disabilities of a retired population, male and female, and recommend one of the places that you studied for each of the most common ailments that came up in your research. The environment of the place should complement the treatment of the complaint.

Gumshoe Geography © 1996 Zephyr Press, Tucson, AZ

Continental

Your Assignment: The Committee is in association with a certain government that would like to gain control of the exports and imports in the following cities. Your job is to visit the sites and make a report about the type and volume of imports and exports that go through each of these cities. Be prepared to assess their value to the world market and to determine the share of the world market that those imports and exports hold. Follow all orders and good luck.

5°19'N	4°02'W	30°03'N	31°15'E
35°55'S	18°22'E	33°37'N	7°35'W
14°40'N	17°26'W	6°30'N	3°30'E
8°50'S	13°15'E	26°00'S	32°30'E
2°03'N	45°22'E	32°54'N	13°11'E

Special Instructions: The Committee has been observing your work and has confidence in your abilities and potential. We would like to give you an opportunity for advancement. The region that you are visiting is often torn by conflict.

- Please find out which sites are in a state of conflict and which are stable at this time.

- Please propose solutions to two of the conflicts that interest you and that you have thoroughly researched.

Continental

Your Assignment: The Committee is helping a major international conglomerate locate a new industrial complex. Your job is to visit the following sites and find out what industries already exist in these places. Be prepared to let us know how healthy these industries are, what sources of power and raw materials feed them, and whether they require a major work force. Follow all orders and good luck.

61°13′N	149°53′W	42°21′N	71°04′W
44°39′N	63°36′W	23°08′N	82°22′W
29°46′N	95°22′W	38°40′N	76°14′W
46°49′N	71°13′W	37°48′N	122°24′W
32°04′N	81°05′W	47°34′N	52°43′W
49°16′N	123°07′W	19°12′N	96°08′W

Special Instructions: The Committee has been observing your work and has confidence in your abilities and potential. We would like to give you an opportunity for advancement. These cities are major hubs for industry and commerce and have rich historical heritages.

- Please research and report back to us about each site's history, letting us know how the sites came to be the way they are.

   Gumshoe Geography © 1996 Zephyr Press, Tucson, AZ

Continental

Your Assignment: The Committee is studying the effects of elevation on a city's standard of living and way of life. Your job is to visit the following sites, find out their elevations, and then research the standards of living of the people at each site. Be prepared to discuss your findings about the effects of elevation on performance at certain selected jobs and careers. Follow all orders and good luck.

39°06'N 84°31'W	64°04'N 139°25'W
39°05'N 108°33'W	20°40'N 103°20'W
35°58'N 83°56'W	48°14'N 101°18'W
46°06'N 64°07'W	25°40'N 100°19'W
28°32'N 81°23'W	52°07'N 106°38'W
47°40'N 117°23'W	46°30'N 81°00'W

Special Instructions: The Committee has been observing your work and has confidence in your abilities and potential. We would like to give you an opportunity for advancement. As a rule, major cities make themselves distinctive by providing their citizens with certain forms of entertainment.

- Please find out about sports teams from these cities and how they do within their own spheres.
- Determine how the team reflects the assumed character of the city.

Continental

Your Assignment: The Committee is assembling a catalogue of environmentally matched sites in the world so that we can relocate groups of people with minimal distress. Your job is to visit the following sites and assess the local environment, climate, and topography and then suggest a site in the United States that would closely match those factors. Be prepared to justify your suggestions. Follow all orders and good luck.

23°39'S	70°42'W	23°16'S	57°40'W
3°27'N	76°31'W	20°27'S	60°50'W
13°31'S	71°59'W	2°10'S	79°50'W
9°45'N	63°11'W	32°54'S	68°50'W
34°53'S	56°11'W	17°48'S	63°10'W

Special Instructions: The Committee has been observing your work and has confidence in your abilities and potential. We would like to give you an opportunity for advancement. The region that you are visiting is rich in local culture.

- Please find out about the culture at the sites that you are visiting and report on the music that is used at local celebrations during the year.

- Please let us know about the celebrations and about the instruments that are used to make the music.

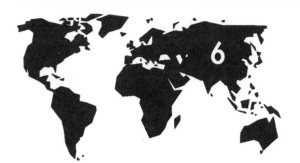

Continental

Your Assignment: The Committee is still in the process of matching environmentally similar sites in the world for low impact relocation of personnel. Your job is to visit the following sites and assess the environmental factors, including climate, topography, and natural environment of the region surrounding the site. Be prepared to recommend matching sites on the continent of Eurasia and also be prepared to justify your recommendations. Follow all orders and good luck.

1°15′S	78°37′W	10°59′N	74°48′W
8°17′S	35°58′W	23°25′S	51°17′W
20°13′S	70°10′W	3°50′S	73°15′W
19°35′S	65°45′W	32°57′S	60°40′W
31°23′S	57°58′W	10°11′N	67°45′W

Special Instructions: The Committee has been observing your work and has confidence in your abilities and potential. We would like to give you an opportunity for advancement. The family culture of this region makes an interesting study.

- Please create a simple profile of a typical native family from each of these sites and tell us what they eat for breakfast, lunch, and dinner.
- Please tell us if they harvest their own food and how they prepare it.

Continental

Your Assignment: The Committee is involved in a transitional government in a certain country and we need some specific historical data. Your job is to visit the following sites and let us know if these countries were ever ruled by a monarchy. Be prepared to tell us about the last time that the monarch was in power and what happened to change the situation. If the monarch is still present, let us know why and how monarchy is working. Follow all orders and good luck.

57°03′N	9°56′E	53°20′N	6°15′W
48°43′N	21°15′E	51°24′N	16°13′E
36°43′N	4°25′W	48°41′N	6°12′E
38°07′N	13°22′E	48°46′N	9°11′E
40°38′N	22°56′E	45°48′N	16°00′E

Special Instructions: The Committee has been observing your work and has confidence in your abilities and potential. We would like to give you an opportunity for advancement. The languages of this region vary widely and come from a host of different roots.

- Please tell us what language the people of these regions speak and what the official language is.
- Find out for us what the roots of the written and spoken local languages are.

   Gumshoe Geography © 1996 Zephyr Press, Tucson, AZ

Continental

Your Assignment: The Committee is helping a new government through the maze of alliances and alignments in this region. Your job is to visit the following sites and report on the alignments and alliance organizations to which the government of each site belongs. Be prepared to discuss the purpose of each organization and name its members. Follow all orders and good luck.

44°50′N	20°30′E	52°31′N	13°24′E
37°53′N	4°46′W	55°53′N	4°15′W
38°43′N	9°08′W	48°52′N	2°20′E
42°45′N	12°29′E	59°20′N	18°03′E
48°12′N	16°22′E	47°20′N	8°35′E

Special Instructions: The Committee has been observing your work and has confidence in your abilities and potential. We would like to give you an opportunity for advancement. During your travels across this region, you have encountered various cultures and ways of life. If you could characterize each of the sites by one thing that it produced, what would that product be?

- Please decide on one product per site and prepare an attractive presentation to let us know of your choices.

Continental

Your Assignment: The Committee is helping prepare a world court lawsuit for a certain nation. Your job is to visit the following sites and find evidence of foreign domination and the slave trade. Be prepared to discuss the effects of this domination on the original population and how the effects are still evident in the local cultures. Follow all orders and good luck.

13°06′N	59°37′W	9°22′N	79°54′W
15°47′N	86°50′W	12°35′N	86°35′W
18°30′N	77°55′W	15°43′N	88°36′W
16°07′N	88°48′W	9°58′N	84°50′W
22°24′N	79°58′W	19°27′N	70°42′W
17°06′N	61°51′W	12°06′N	68°56′W

Special Instructions: The Committee has been observing your work and has confidence in your abilities and potential. We would like to give you an opportunity for advancement. Few places in this region are far from the sea or the rain forest.

- Please find out for us how the people use the resources found in the sea or the rain forest.

- Find out what they harvest and how; find out what the market is for the harvest and who benefits.

Continental

Your Assignment: The Committee is studying how nations go through periods of adjustment after becoming independent from outside control. Your job is to visit the following sites and find out if their respective governments have gone through such transitions. Be prepared to discuss the types of conflict that arise from the transitions and the conflicts that are actually happening in that region. Follow all orders and good luck.

17°15′N	88°46′W	19°45′N	72°15′W
8°25′N	82°27′W	18°00′N	76°50′W
10°00′N	83°02′W	12°53′N	85°57′W
16°14′N	61°23′W	18°01′N	66°37′W
14°50′N	91°31′W	13°29′N	88°11′W
20°01′N	75°49′W	12°03′N	61°45′W

Special Instructions: The Committee has been observing your work and has confidence in your abilities and potential. We would like to give you an opportunity for advancement. This region is a great place to vacation and the whole world knows it.

- Please find out how the local population views the influx of outsiders and how they deal with the outsiders, whether they are visiting for business or pleasure.

Continental

Your Assignment: The Committee is looking at the relationship between standard of living and natural environment. Your job is to visit the following sites and report on the natural environment of the region around each one. Be prepared to discuss the standard of living in each of the sites and how it is characterized. Try to find correlations and be able to demonstrate them. Follow all orders and good luck.

30°10'N	48°50'E	18°58'N	72°50'E
22°15'N	114°10'E	14°35'N	121°00'E
35°06'N	129°03'E	16°47'N	96°10'E
31°41'N	121°28'E	1°17'N	103°51'E
35°40'N´	139°46'E	43°10'N	131°56'E

Special Instructions: The Committee has been observing your work and has confidence in your abilities and potential. We would like to give you an opportunity for advancement. Some of the great religions of the world are centered in the sites that you are visiting.

- Please find out about the major religions in each site and report on the major beliefs and celebrations that characterize each religion.

- Please let us know how each religious group tolerates other religious groups in the region.

Gumshoe Geography © 1996 Zephyr Press, Tucson, AZ

Continental

Your Assignment: The Committee is helping a certain international organization update the data on endangered species in the world. Your job is to visit the following sites and find out about the status and condition of any endangered populations of animals. Be prepared to discuss what those nations are doing to preserve the populations and how efforts from the outside are handled in those countries. Follow all orders and good luck.

6°10′S	106°46′E	34°31′N	69°12′E
3°10′N	101°42′E	35°00′N	135°45′E
28°36′N	77°12′E	11°33′N	104°55′E
37°34′N	127°00′E	35°40′N	51°26′E
47°55′N	106°53′E	30°30′N	114°20′E

Special Instructions: The Committee has been observing your work and has confidence in your abilities and potential. We would like to give you an opportunity for advancement. Various cultures have different ways of thinking about the basic family structure and you can learn from your visits to these sites.

- Please find out about and report on how people in the regions around these sites think about their families and how they raise their children, both males and females.

- Find out what status children have in the family unit and at what ages they experience transitions.

5
URBAN TOPICS

World Map Activity

Urban Topics

Your Assignment: The Committee has recently acquired a major manufacturing firm with several established international markets. We want to expand the number of markets to increase the access of The Committee to world commerce. Your job is to visit and observe the following port cities and to analyze them for possible inclusion in our network. Follow all instructions and good luck.

12°28′S	130°50′E	44°39′N	63°36′W
24°10′N	110°18′W	10°39′N	61°31′W
23°39′S	70°24′W	14°40′N	17°26′W
4°03′S	39°31′E	51°54′N	8°28′W
39°36′N	19°55′E	24°52′N	67°03′E
20°52′N	106°41′E	41°45′N	140°43′E

Special Instructions: The Committee has been observing your work and has confidence in your abilities and potential. We would like to give you an opportunity for advancement.

- Please do some research and report to us about where in the world we could have our ships built.

- Decide which characteristics of a ship-building port are important and then decide which locations would be in the best interest of The Committee and explain why.

Gumshoe Geography © 1996 Zephyr Press, Tucson, AZ

Urban Topics

Your Assignment: The Committee has a project in motion that involves the highest levels of government and certain vital natural resources. Your job is to visit the following sites and investigate the nature and volume of shipping in and out of these ports. Be prepared to discuss the similarities and differences of these sites. Find out about their dependence on certain raw materials and report back to us. Follow all orders and good luck.

35°36'N	140°07'E	29°46'N	95°22'W
35°32'N	139°43'E	34°41'N	135°10'E
43°18'N	5°24'E	29°04'N	48°09'E
29°58'N	90°07'W	40°43'N	74°01'W
51°55'N	4°28'E	35°27'N	139°39'E

Special Instructions: The Committee has been observing your work and has confidence in your abilities and potential. We would like to give you an opportunity for advancement. The sites that you are to visit are for the most part bustling port cities with nothing to attract people except commerce.

- Please find out if they have developed art or music communities.
- Report back to us about museums there and whether they have resident symphonies, opera companies, or ballet companies.

Urban Topics

Your Assignment: The Committee has been investigating certain places in the world because of their reputation as "forbidden cities." Your job is to visit the following sites and find out what it is about them that makes them forbidden. Be prepared to discuss the culture that exists in the sites and, if their reputation existed in the past, look into the history of the places and report why they were called forbidden. Make a detailed report on the one that you find most interesting. Follow all orders and good luck.

29°42'N	91°07'E	21°27'N	39°49'E
40°15'N	24°15'E	11°33'N	92°15'E
39°55'N	116°23'W	52°38'N	6°14'E

Special Instructions: The Committee has been observing your work and has confidence in your abilities and potential. We would like to give you an opportunity for advancement. These sites held great interest for the people who explored or discovered them.

- Please research the cities and then compose a fictional account of life in one of the cities that you find interesting. Include as much factual information as you can, but weave it into a convincing story that we all might enjoy and that might illuminate our understanding of the place.

Urban Topics

Your Assignment: The Committee has identified the following sites as potentially influential in world affairs. Your job is to visit these sites and find out what they have in common and why we chose them. Be prepared to find patterns that are characteristic of these types of cities and report on those patterns. Follow all orders and good luck.

18°58′N	72°50′E	34°36′S	58°27′W
22°32′N	88°22′E	34°03′N	118°15′W
19°24′N	99°09′W	55°45′N	37°35′E
40°43′N	74°01′W	35°37′N	137°14′E
22°54′S	43°15′W	23°32′S	46°37′W
37°34′N	127°00′E	35°40′N	139°46′E

Special Instructions: The Committee has been observing your work and has confidence in your abilities and potential. We would like to give you an opportunity for advancement. These sites are all major centers for everything from commerce to the arts.

- Please investigate life in these places and try to find cultural patterns that are able to survive the crunch of metropolitan life.

- Report on those patterns and let us know how we can help preserve them in their present setting.

Urban Topics

Your Assignment: The Committee has been helping a major international firm locate suitable sites, with distinctive cultural character, for the location of their facilities. Your job is to visit the following sites and find out about their most prominent characteristics. Be prepared to discuss the reputation of people from those cities, whether positive or negative. Follow all orders and good luck.

30°03′N	31°15′E	28°40′N	77°13′E
51°27′N	7°01′E	6°10′S	106°46′E
24°52′N	67°03′E	6°27′N	3°23′E
12°03′S	77°03′W	51°30′N	0°10′W
14°35′N	121°00′E	48°52′N	2°20′E
31°41′N	121°28′E	35°40′N	51°26′E

Special Instructions: The Committee has been observing your work and has confidence in your abilities and potential. We would like to give you an opportunity for advancement. Large metropolitan areas are usually lacking open spaces, and the people must work hard to preserve what park space they have. Yet the areas are not free of wildlife; certain species thrive in urban areas.

- Please find out about animals that live in these sites and report back on the different kinds.

- If there are patterns, please let us know about them.

Gumshoe Geography © 1996 Zephyr Press, Tucson, AZ

Urban Topics

Your Assignment: The Committee has been asked to help in choosing sites for future summer Olympic Games. Your job is to visit the following sites and find out what they have in common. Be prepared to talk about why the sites were selected for the Games and let us know what we should be looking for in a future Olympic site. Follow all orders and good luck.

37°59'N	23°44'E	41°23'N	2°11'E
52°31'N	13°24'E	60°01'N	24°58'E
51°30'N	0°10'W	37°49'S	144°58'E
45°31'N	73°34'W	48°09'N	11°35'E
48°52'N	2°20'E	42°45'N	12°29'E
37°34'N	127°00'E	35°40'N	139°46'E

Special Instructions: The Committee has been observing your work and has confidence in your abilities and potential. We would like to give you an opportunity for advancement. The Olympics have changed over the years in many ways.

- Please put the sites in chronological order according to when they hosted the Games, and find out how the Games changed each time they were held.

- Tell us what stood out and the innovations that were the hallmark of each site.

Urban Topics

Your Assignment: The Committee has been asked to help potential host sites make models for facilities to be considered by the Olympic Committee. Your job is to visit the following sites and assess how well the facilities for the Games fit in with the surroundings and the local environment. Be prepared to discuss any innovations in placement or construction of facilities that have happened from one Games to the next. Let us know how the hosts use the facilities after the Games are finished at that site. Follow all orders and good luck.

51°03′N	114°05′W	45°55′N	6°52′E
46°32′N	12°08′E	47°30′N	11°06′E
45°10′N	5°43′E	47°16′N	11°24′E
44°18′N	73°59′W	59°55′N	10°45′E
43°03′N	141°21′E	43°50′N	18°25′E
36°45′N	119°45′W	46°32′N	9°49′E

Special Instructions: The Committee has been observing your work and has confidence in your abilities and potential. We would like to give you an opportunity for advancement. All winter Olympic sites depend on one completely undependable factor: snow.

- Please find out about each of the Games in the sites that are listed and let us know if each had enough snow and how the hosts handled the lack of snow when it occurred.

- If there are ways to predict if a site will be a good one, let us know how that is done.

 Gumshoe Geography © 1996 Zephyr Press, Tucson, AZ

Urban Topics

Your Assignment: The Committee is in the process of making a catalogue of the capital cities of the world. Your job is to visit the following sites and determine the stability of the government in power at the sites. Be prepared to tell us about any changes that have taken place in those governments in recent history and how they have taken place. Let us know how you feel about the stability of each of the governments. Follow all orders and good luck.

39°56'N	32°52'E	40°23'N	49°51'E
60°01'N	24°58'E	6°10'S	106°46'E
34°31'N	69°12'E	15°50'N	33°00'E
8°50'S	13°15'E	6°19'N	10°48'W
45°25'N	75°42'W	64°09'N	21°57'W
37°34'N	127°00'E	52°15'N	21°00'E

Special Instructions: The Committee has been observing your work and has confidence in your abilities and potential. We would like to give you an opportunity for advancement. The sites that you are visiting are true seats of power and we would like for you to find out about the people that wield that power.

- Please find out about the leaders of these nations and report on the most famous or infamous leader of each of these countries, present or past.

- Please write a detailed report about the public life of the leader of your choice.

Urban Topics

Your Assignment: The Committee needs information on the governments in the following capitals in order to determine where to establish new subcommittees. Your job is to visit these capitals and, in addition to reporting the regular information, note the forms of government and important names and dates. We trust that your report will be complete. Follow all instructions and good luck.

15°47′S	47°55′W	50°50′N	4°20′E
34°36′S	58°27′W	35°17′S	149°08′E
23°08′N	82°22′W	14°35′N	121°00′E
55°45′N	37°35′E	45°25′N	75°42′W
39°55′N	116°23′E	59°20′N	18°03′E
35°40′N	51°26′E	41°17′S	174°46′E

Special Instructions: The Committee has been observing your work and has confidence in your abilities and potential. We would like to give you an opportunity for advancement.

- Please predict how each government would stand on three current world issues and explain your answers in light of current political conditions in the designated countries.

    Gumshoe Geography © 1996 Zephyr Press, Tucson, AZ

Urban Topics

Your Assignment: The Committee has been asked to help an emerging nation design a capitol that will reflect what it stands for. Your job is to visit the following sites and find out about the ideas that went into designing the capitol sections of the cities. Be prepared to discuss the character of the architecture and the prominence of the capitol buildings and how they reflect the character of the culture of the city and of the nation. Follow all orders and good luck.

9°10'N	7°11'E	31°57'N	35°56'E
48°09'N	17°07'E	13°06'N	59°37'W
53°20'N	6°15'W	15°25'S	28°17'E
53°45'N	27°34'E	39°01'N	125°45'E
0°13'S	78°30'W	16°47'N	96°10'E
42°45'N	12°29'E	34°00'N	9°00'E
38°54'N	77°01'W		

Special Instructions: The Committee has been observing your work and has confidence in your abilities and potential. We would like to give you an opportunity for advancement. Capitol cities are places that are usually dedicated not only to the preservation of the present state of the nation but also to the remembrance of the nation's past glory.

- Please find out about any monuments that can be found in the sites and tell us what or whom they honor.

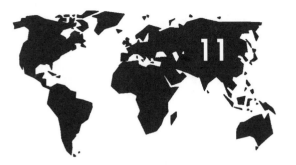

Urban Topics

Your Assignment: The Committee has been considering forming a small national state. Your job is to visit the following sites and find out about the ways that these small countries are governed. Be prepared to discuss the reasons that they have been able to exist for so long and not be swallowed up by larger neighbors. Follow all orders and good luck.

42°30'N	1°30'E	17°03'N	61°48'W
12°10'S	44°10'E	12°07'N	61°40'W
47°10'N	9°30'E	3°15'N	73°00'E
35°50'N	14°30'E	43°42'N	7°23'E
43°55'N	12°28'E	1°17'N	103°51'E

Special Instructions: The Committee has been observing your work and has confidence in your abilities and potential. We would like to give you an opportunity for advancement. These countries are so small that we assume that it would be difficult to form a national character and culture.

- Please investigate this theory and report back to us on any forms of culture or distinction that you observe in these places.

- Please try to find out how the people see themselves as a nation and as players on the world stage.

Gumshoe Geography © 1996 Zephyr Press, Tucson, AZ

6
RURAL CULTURE

World Map Activity

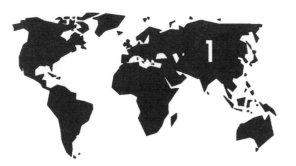

Rural Culture

Your Assignment: The Committee wants to study the economics of selected rural areas in the Western Hemisphere. Your job is to visit the following locations, even though there may not be a major city nearby, find a group of local people in each place, and spend some time making environmental and cultural observations. You will be asked to report on the economics of each location. Follow all instructions and good luck.

60°N	155°W	65°N	20°W
40°N	100°W	30°N	85°W
30°N	115°W	15°N	85°W
20°N	75°W	30°N	70°W
5°S	65°W	45°S	60°W

Special Instructions: The Committee has been observing your work and has confidence in your abilities and potential. We would like to give you an opportunity for advancement.

■ Please speculate about changes in the community brought on by influences from the outside world. Which specific change would most improve and which would do the most damage to the economic health and well-being of each location?

    Gumshoe Geography © 1996 Zephyr Press, Tucson, AZ

Rural Culture

Your Assignment: The Committee has been searching for a rural culture within which to implement some of our social change models. Your job is to visit the following sites and assess the local cultures. Be prepared to tell us how close the present culture is to its roots and how and why it has changed. We also want to know how much Western influence there is at these sites. Follow all orders and good luck.

44°S	170°E	20°N	106°E
41°N	141°E	30°N	105°E
32°N	92°E	6°N	81°E
35°N	70°E	15°S	28°E
36°N	6°E	50°N	2°E
66°N	18°E		

Special Instructions: The Committee has been observing your work and has confidence in your abilities and potential. We would like to give you an opportunity for advancement. In many cultures the people appear to welcome tourists, but in reality they shun outsiders.

- Please find out how the people in these sites view other people.
- Find out what methods they use to expel outsiders, turn them away, welcome them, or just ignore them.

World Map Activity

Rural Culture

Your Assignment: The Committee is looking for a native culture that is intact and relatively unchanged from its early roots. It hopes to conduct studies about the effects of environmental changes on ways of life in primary cultures. Your job is to visit the locations listed below, make observations, and file your regular report. Then choose the group most likely to be successful in our study and write a narrative of your stay in their culture. We trust that you will be thorough with your study, including names and family structures and interpersonal relationships in your report. Also include information on specific plants and animals that each tribe or group has domesticated and how the tribe uses these resources. Follow all instructions and good luck.

70°N	160°W		35°N	110°W
20°N	90°W		30°S	70°W
5°S	60°W		5°S	35°E
20°N	50°E		70°N	25°E
18°S	178°E		45°N	105°E
25°S	125°E		0°	20°E

Special Instructions: The Committee has been observing your work and has confidence in your abilities and potential. We would like to give you an opportunity for advancement.

- Please make a model or a detailed drawing of a typical village for the group that you have chosen, or illustrate and describe the food web and water source in the group's environment.

4

Rural Culture

Your Assignment: The Committee would like you to choose two strategic locations in each of the six continents that have native populations so that we may implement a worldwide communication network. The choice of locations is up to you, but we want you to justify your choices. Time is short, so make the first choices the best ones. Follow all instructions and good luck.

Special Instructions: The Committee has been observing your work and has confidence in your abilities and potential. We would like to give you an opportunity for advancement.

- Please do some research on current communication technology and make evaluations as to type, strengths, and weaknesses of the most suitable technology.

- Also, please report on the feasibility of the project. Justify your answers.

7

WORLD CLIMATIC REGIONS

World Map Activity

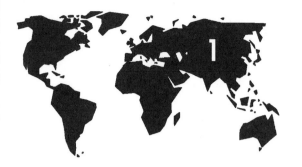

World Climatic Regions

Your Assignment: The Committee is seeking to publish a catalogue of the various ecosystems in the many climatic regions of the world. Your job is to visit the following sites and find out what animals thrive in this type of setting. Be prepared to demonstrate in a graphic format the various food webs that characterize the different sites. Follow all orders and good luck.

18°50'S	62°10'W	0°00'	17°00'E
2°00'N	102°30'E	25°52'N	81°23'W
36°30'N	76°30'W	30°50'N	47°10'E
37°00'N	6°15'W	45°30'N	29°45'E
4°50'N	6°00'E	30°42'N	82°20'W
18°00'S	56°00'W	19°00'S	52°00'W
3°00'N	114°00'E	5°24'N	0°20'W

Special Instructions: The Committee has been observing your work and has confidence in your abilities and potential. We would like to give you an opportunity for advancement. Such regions are often inhospitable to human settlement.

- Please find out for what this type of region is useful and for what it has been used in the past.

- If there is folklore associated with any of the sites, please let us know about it.

World Climatic Regions

Your Assignment: The Committee is interested in the development of certain pharmaceutical companies and in helping them find new sources for new medicines. Your job is to visit the following sites and report on any medicines that have already been discovered from among the plant species in this type of region. Be prepared to make suggestions as to how to expedite the discovery of more medicinal plants. Follow all orders and good luck.

17°15′N	88°46′W	23°43′N	90°25′E
3°43′S	38°30′W	20°52′N	106°41′E
4°22′N	7°43′W	2°32′S	140°42′E
0°25′N	25°12′E	16°42′N	74°13′E
3°10′N	101°42′E	3°08′S	60°01′W
8°58′N	79°31′W	12°59′S	38°31′W

Special Instructions: The Committee has been observing your work and has confidence in your abilities and potential. We would like to give you an opportunity for advancement. This region of the Earth is being cut away and it cannot replace itself.

- Please find out about the people who are cutting down the rain forests and find out why they are doing it.

- Please propose measures to help them with their needs and to save this valuable resource.

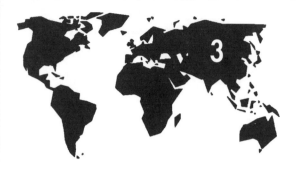

World Climatic Regions

Your Assignment: The Committee has just gained interest in a northern oil development firm. We are looking at running a pipeline across a very delicate environment. Your job is to visit the following sites and find out about the environment in each site. Be prepared to discuss the way in which the ecosystem works and how we can best preserve it while using it for our purposes. Follow all orders and good luck.

64°34′N	40°32′E	64°51′N	147°43′W
51°17′N	80°39′E	53°00′N	150°00′E
65°00′N	27°00′E	54°54′N	69°06′E
47°34′N	52°43′W	59°55′N	30°15′E
63°50′N	20°15′E	62°27′N	114°21′W

Special Instructions: The Committee has been observing your work and has confidence in your abilities and potential. We would like to give you an opportunity for advancement. People who live in this region have limited opportunities to earn a living.

- Please find out about life in the region and report to us about what people do, how families are structured, what the problems are that the people experience, and what is attractive about life in this region.

Gumshoe Geography © 1996 Zephyr Press, Tucson, AZ

World Climatic Regions

Your Assignment: The Committee would like to become involved in the management of the supply of renewable resources in the world. Your job is to visit the following sites and find out about the region and what resources it has to offer. Be prepared to discuss efforts to increase the yield and general health of the resources and tell us how government plays a part in their preservation at each site. Follow all orders and good luck.

44°49′N	68°47′W	45°34′S	72°04′W
47°59′N	122°13′W	55°53′N	4°15′W
54°17′N	31°00′E	51°20′N	12°20′E
38°03′N	84°30′W	43°03′N	141°21′E
37°50′N	112°37′E	44°24′S	171°15′E

Special Instructions: The Committee has been observing your work and has confidence in your abilities and potential. We would like to give you an opportunity for advancement. This region is full of recreational opportunities.

- Please find out about such opportunities near your sites and let us know what the sites have to offer and to whom.

- Please also propose other recreational sites that could be developed without hurting the surrounding environment.

World Climatic Regions

Your Assignment: The Committee is studying the possibility of establishing a few secret bases in this region. Your job is to visit the following sites and report to us about the natural environment and what resources we would find attractive. Be prepared to tell us about seasonal changes that take place and to make recommendations as to what the area might be good for and how we should construct a facility in such a region. Follow all orders and good luck.

36°52′S	174°45′E	35°17′S	149°08′E
23°07′N	113°18′E	25°25′S	49°15′W
42°53′S	147°19′E	28°03′N	81°57′W
32°50′N	83°38′W	26°18′S	31°07′E
32°47′N	129°56′E	31°25′S	26°52′E

Special Instructions: The Committee has been observing your work and has confidence in your abilities and potential. We would like to give you an opportunity for advancement. This area of the world has become increasingly popular and the population has swelled in many regions.

- Please study the area and report to us about the factors that might influence the great population growth in this region.

- Please suggest measures that should be taken to ensure no harmful effects occur as a result of the growth.

Gumshoe Geography © 1996 Zephyr Press, Tucson, AZ

World Climatic Regions

Your Assignment: The Committee has an agreement with a certain government to build a test facility in one of your destinations. Your job is to visit the following sites and get the information about the land and how it is used. Be prepared to discuss the stability of the local people, how much they depend on the land, and how it might impact the culture to take a large tract of land and use it for other purposes. Follow all orders and good luck.

4°22'N	18°35'E	11°05'S	43°10'W
15°56'S	50°08'W	18°13'S	127°40'E
17°50'S	31°10'E	0°19'N	32°35'E
20°44'S	139°30'E	14°00'N	2°00'E
27°50'S	64°15'W	22°13'N	97°51'W

Special Instructions: The Committee has been observing your work and has confidence in your abilities and potential. We would like to give you an opportunity for advancement. This type of region is always full of animal life no matter where it is found in the world.

- Please take a survey of the different regions and create a graphic display of the food webs that are typical of the regions that you are visiting.

World Climatic Regions

Your Assignment: The Committee is trying to establish an agrarian economy in a small nation of which we have recently gained control. Your job is to visit the following sites and find out about the types of crops that are grown in the region. Be prepared to distinguish between the ones that are historically rooted to the region and the ones that are relatively new to the area. Show us a little about the history of farming in the region. Follow all orders and good luck.

32°42'S	26°20'E	36°47'N	3°30'E
41°38'N	41°38'E	33°53'N	35°30'E
35°55'S	18°22'E	37°55'N	22°53'E
36°48'N	34°38'E	33°19'N	115°38'E
40°41'N	14°47'E	34°03'N	118°15'W
37°23'N	5°59'W	33°02'S	71°38'W

Special Instructions: The Committee has been observing your work and has confidence in your abilities and potential. We would like to give you an opportunity for advancement. This area of the world is very conducive to the transaction of all types of business.

- Please tell us about commerce in the regions, from the one-person business to the multinational corporations.

- Tell us how they work and if there are any common patterns in the ways people do business in this region.

Gumshoe Geography © 1996 Zephyr Press, Tucson, AZ

World Climatic Regions

Your Assignment: The Committee is in the process of cataloguing the dry regions of the world so that they can be categorized for proper and most efficient use. Your job is to visit the following sites and find out how stable the topography of each site is. Be prepared to discuss the movement of the sand in the sites and the encroachment of the desert on arable land. Make a categorized list of the sites using criteria that you choose. Follow all orders and good luck.

28°00'N	32°00'E	22°30'S	69°15'W
28°30'N	106°00'W	36°30'N	117°00'W
24°30'S	126°00'E	43°00'N	106°00'E
33°00'N	57°00'E	23°00'S	15°00'E
36°00'N	111°20'W	21°00'N	6°00'E
32°00'N	40°00'E	27°00'N	70°00'E

Special Instructions: The Committee has been observing your work and has confidence in your abilities and potential. We would like to give you an opportunity for advancement. Travel in this type of region is a challenge.

- We would like you to write a travel manual for our members who have to be here.

- Please suggest how to make our members' travel in the region safe and comfortable.

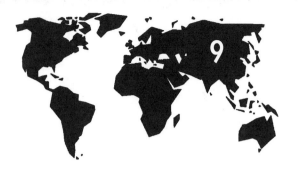

World Climatic Regions

Your Assignment: The Committee has been studying the possibility of opening new sources of mineral resources in your region. Your job is to visit the following sites and find out about resources that might be available for mining. Be prepared to discuss the technology that would be required on a venture such as this one. Make recommendations as to the best way for us to proceed. Follow all orders and good luck.

21°30'S	125°00'E	28°30'S	127°45'E
23°00'S	22°00'E	39°00'N	60°00'E
42°00'N	64°00'E	24°00'N	25°00'E
35°00'N	117°00'W	20°30'N	33°00'E
18°00'N	45°00'E	37°59'N	120°23'W
39°00'N	83°00'E		

Special Instructions: The Committee has been observing your work and has confidence in your abilities and potential. We would like to give you an opportunity for advancement. In this harsh environment many things have to be considered in order to create hospitable shelter.

- Please report on the ways that the local people meet their needs for shelter and then make some proposals of your own using the best of what you have already seen.

 Gumshoe Geography © 1996 Zephyr Press, Tucson, AZ

World Climatic Regions

Your Assignment: The Committee knows the importance of grain to the world and is aware of the hunger problem. The Committee would like to offer its assistance. Your job is to visit the following sites and find out what species of grains people are growing in each, how it is grown and harvested, what people are doing to enhance the crop, and how healthy the soil is for the longevity of the crop. Be prepared to discuss innovations that might be able to be used throughout the world. Follow all orders and good luck.

39°56′N	32°52′E	29°00′S	58°00′W
48°27′N	34°59′E	49°33′N	106°21′E
32°45′N	97°20′W	24°40′S	25°55′E
30°59′S	150°15′E	43°50′N	73°10′E
50°25′N	104°39′W	29°41′S	53°48′W
43°32′N	96°44′W		

Special Instructions: The Committee has been observing your work and has confidence in your abilities and potential. We would like to give you an opportunity for advancement. In order to harvest large crops in such a fertile region, various farm methods are called for.

- Please find out about traditional and modern farm methods in this region.

- Let us know about any new advances or trends in machinery, rotation, irrigation, pesticides, and so on.

8

SPECIFIC PHYSICAL GEOGRAPHY

World Map Activity

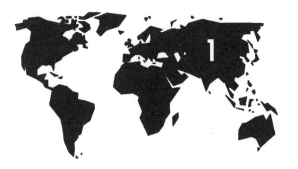

Specific Physical Geography

Your Assignment: The Committee would like to establish a secret communications headquarters and supply depot. The facility needs to be on an island. Your job is to visit the following islands or island groups and make observations. You may also choose three other locations that we have not specified. We trust that you will be careful. Follow all instructions and good luck.

50°N	3°W	36°N	28°E
4°N	73°30'E	10°S	160°E
22°N	158°W	54°S	37°W
13°N	59°30'W	33°30'S	78°30'W
63°30'N	170°30'W	47°N	63°W
29°N	119°W	32°45'N	17°W

Special Instructions: The Committee has been observing your work and has confidence in your abilities and potential. We would like to give you an opportunity for advancement.

- Please make topographical drawings of the three islands that you consider to be the best choices for our facility.

- Show high points of land, natural harbors, reefs and other obstructions, approaches, and any other features that you think are important.

Specific Physical Geography

Your Assignment: The Committee has been involved in doing atmospheric studies and we are looking for new sites for our studies. Your job is to visit the following mountain peaks and find out how accessible they are and how stable the weather patterns around them are. Be prepared to discuss the ease or difficulty in reaching them and what equipment we would need if we were to explore them. Follow all orders and good luck.

32°39'S	70°00'W	45°50'N	6°52'E
10°50'N	73°45'W	43°21'N	42°26'E
27°59'N	86°56'E	35°53'N	76°30'E
3°04'S	37°22'E	0°10'S	37°20'E
27°08'S	109°26'W	63°30'N	151°00'W
19°02'N	98°38'W	31°03'N	7°55'W
78°35'S	85°25'W	1°45'S	133°25'E

Special Instructions: The Committee has been observing your work and has confidence in your abilities and potential. We would like to give you an opportunity for advancement. The people who live in the regions surrounding these peaks usually have stories associated with the peaks.

- Please visit the regions and find out how the local population feels about the peaks and about outsiders being near them.

Specific Physical Geography

Your Assignment: The Committee has been compiling a catalogue of specific strategic sites in the world and it is almost complete. Your job is to visit the following mountain passes and learn about the geography of each site. Be prepared to discuss how it could be strategic in relation to certain activities. Find out about those activities and also determine how accessible the passes are. Follow all orders and good luck.

20°23′N	18°18′E	34°50′N	67°50′E
47°45′N	7°0′E	47°00′N	11°30′E
36°51′N	83°19′E	39°19′N	120°20′W
45°50′N	8°10′E	34°05′N	71°10′E
40°24′N	105°05′W	38°31′N	73°41′E
43°01′N	1°19′W	42°22′N	108°55′W
46°30′N	8°30′E	38°48′N	22°32′E

Special Instructions: The Committee has been observing your work and has confidence in your abilities and potential. We would like to give you an opportunity for advancement. The locations that you have been exploring have been historically important for many reasons.

- Please investigate their importance and report back to us about their historical significance.
- Also let us know if they are still important and why or why not.

Specific Physical Geography

Your Assignment: The Committee has been working with a multinational alliance to head off a new round of nuclear testing. Your job is to choose, locate, and visit five atolls and five reefs. The South Pacific Ocean is a good place to begin looking. Report on the ecosystem of the sites. Be prepared to discuss the dangers of using the sites for weapons testing. Use any evidence that you can find. Follow all orders and good luck.

Special Instructions: The Committee has been observing your work and has confidence in your abilities and potential. We would like to give you an opportunity for advancement. The regions that have been explored are death traps for some people and gold mines for others.

- Please find out about the legacy of shipwrecks and treasure hunting at these sites and let us know about accounts of great finds and great losses.

Specific Physical Geography

Your Assignment: The Committee is interested in the supply of fresh water in the world and in gaining control in regions where there is potential for new sources. Your job is to choose, locate, and visit five glaciers and find out about the size and volume of the ice sheets. Be prepared to discuss the feasibility of using the glaciers as sources of fresh water for nations that are in need of it, and propose a system for making it happen. Follow all orders and good luck.

Special Instructions: The Committee has been observing your work and has confidence in your abilities and potential. We would like to give you an opportunity for advancement. Exploration of these regions is full of dangers.

- We would like you to compile a list of precautions for our members to observe.

- Please list the dangers and propose ways of making a journey to the sites as safe as possible.

Gumshoe Geography © 1996 Zephyr Press, Tucson, AZ

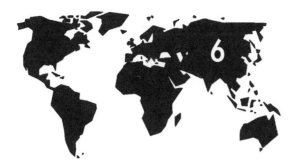

Specific Physical Geography

Your Assignment: The Committee has been asked to investigate the possibility of establishing fresh-water commercial fisheries in certain regions of the world. Your job is to visit the following lakes and find out about the geography of the sites. Determine the feasibility of establishing a fishery in each site and be prepared to recommend a procedure for us to follow to get the job done. Follow all orders and good luck.

45°00'N	60°00'E	53°00'N	107°40'E
42°00'N	50°30'E	42°08'N	80°04'W
64°54'N	125°35'W	61°30'N	114°00'W
48°00'N	88°00'W	12°00'S	34°30'E
6°00'S.	29°30'E	15°50'S	69°20'W
58°88'N	13°30'E	1°00'S	33°00'E

Special Instructions: The Committee has been observing your work and has confidence in your abilities and potential. We would like to give you an opportunity for advancement. These lakes are some of the largest in the world.

- We would like you to find out about the people who live near them.
- Please find out to what extent those people depend on the lakes for their livelihood.
- Find out how each lake affects the family and community structures at the sites.

Specific Physical Geography

Your Assignment: The Committee is involved in the regulation of world shipping and is examining certain sites for the establishment of facilities for control of that industry. Your job is to visit the following straits and tell us about the geography of the region. Be prepared to discuss the use of the strait as a shipping lane and about the possibility of controlling the area. Let us know if you think the governments that are in control of the land would be amenable to our facilities. Follow all orders and good luck.

39°00'S	145°00'E	65°30'N	169°00'W
41°00'N	29°00'E	34°40'N	129°00'E
41°18'N	9°15'E	24°00'N	81°00'W
35°57'N	5°36'W	26°43'N	56°15'E
54°00'S	71°00'W	2°30'N	101°20'E
24°00'N	119°00'E	50°00'N	141°15'E

Special Instructions: The Committee has been observing your work and has confidence in your abilities and potential. We would like to give you an opportunity for advancement. Commerce in the world depends a great deal on the health of the shipping industry.

- Please investigate the life of a typical merchant marine and let us know what training is necessary, what the job requirements are, who does what on a voyage, and how lucrative the job is.

   Gumshoe Geography © 1996 Zephyr Press, Tucson, AZ

Specific Physical Geography

Your Assignment: The Committee has been contacted by a certain international marine agency for our help in re-establishing a network of lighthouses as a safety net if electronics should fail. Your job is to visit the following capes and make a survey of their geographic characteristics. Be prepared to determine if they have ever had lighthouse facilities and whether they would be useful present-day sites for such installations. Follow all orders and good luck.

20°46'N	17°03'W	8°38'N	104°44'E
21°36'N	87°07'W	55°59'S	67°16'W
7°31'S	149°59'E	34°21'S	18°28'E
4°22'N	7°44'W	46°40'N	53°10'W
37°55'N	16°04'E	25°36'S	45°08'E
58°37'N	5°01'W	10°40'S	142°30'E

Special Instructions: The Committee has been observing your work and has confidence in your abilities and potential. We would like to give you an opportunity for advancement. More than any other geographical features, the capes of the world have been given names that imply a colorful history.

- Please find out about the names of the capes on the list and report on any stories or folklore that are associated with the sites.

Specific Physical Geography

Your Assignment: The Committee is involved in the worldwide shipping trade and is looking for ways to make it more efficient and cost effective. Your job is to visit the following capes and study the geography and population of the area. Be prepared to make recommendations as to the feasibility and viability of the construction of offshore cargo ports in these sites to relax some of the overcrowding. Follow all orders and good luck.

71°17′S	170°14′E	35°00′S	136°00′E
8°40′N	77°34′E	5°29′S	35°16′W
40°30′S	172°43′E	18°27′S	12°01′E
11°49′N	51°15′E	35°13′N	75°32′W
50°52′N	156°40′E	40°25′N	124°25′W
21°57′S	43°16′E	14°43′N	17°30′W

Special Instructions: The Committee has been observing your work and has confidence in your abilities and potential. We would like to give you an opportunity for advancement. Most of the capes on the list were named by the seafaring men who had to get around them.

- Please do some research on the names and report any interesting stories to us.

- Please propose new names for the capes based on the present conditions in the area.

Specific Physical Geography

Your Assignment: The Committee is designing a model for a national defense early warning network for our country. Your job is to visit the following sites and, after studying the geography of the sites, find any common characteristics that they might share. Be prepared to profile a fictional site that would be the most typical in terms of natural and human resources and recommend how we could build a facility there. Follow all orders and good luck.

42°39′N	70°38′W	42°50′N	124°34′W
28°30′N	80°35′W	37°17′N	76°00′W
42°52′N	70°22′W	33°53′N	89°32′W
48°32′N	124°43′W	35°13′N	75°32′W
38°56′N	74°54′W	40°25′N	124°25′W
77°35′N	34°40′W	25°12′N	81°05′W
29°40′N	85°22′W		

Special Instructions: The Committee has been observing your work and has confidence in your abilities and potential. We would like to give you an opportunity for advancement. These places that you have visited are part of American history.

- Please find out about their significance and tell us how they were named and why.

- Please prepare a detailed narrative about the naming of the one that most interests you.

Specific Physical Geography

Your Assignment: The Committee is helping an international organization design monitoring equipment in hopes of someday being able to predict volcanic eruptions more effectively. Your job is to visit the following volcanic sites and tell us about the geography of the regions and how they have been impacted by the volcanoes. Be prepared to discuss present methods of prediction and monitoring and let us know about any new technology on the horizon. Follow all orders and good luck.

4°12′N	9°11′E	19°33′N	103°38′W
37°50′N	14°55′E	9°59′N	83°51′W
12°10′S	44°15′E	1°42′S	101°16′E
19°24′N	155°17′W	31°35′N	130°39′E
60°29′N	152°45′W		

Special Instructions: The Committee has been observing your work and has confidence in your abilities and potential. We would like to give you an opportunity for advancement. Volcanoes are fearsome sites, and local populations always live in awareness of their power.

- Please find out how the people of each site handle the insecurity associated with life in proximity to a live or dormant volcano.

Gumshoe Geography © 1996 Zephyr Press, Tucson, AZ

Specific Physical Geography

Your Assignment: The Committee is interested in studying cultures that return to the site of a natural disaster and rebuild even under the threat of another disaster. Your job is to visit the following volcanic sites and report on the last major eruption and on the culture that was in existence when it happened. Be prepared to discuss each culture's return or abandonment of the site and the reasons why they made the choice. Follow all orders and good luck.

6°07′S	105°24′E
37°50′N	14°55′E
46°12′N	122°11′W
40°49′N	14°26′E
8°14′S	117°55′E

Special Instructions: The Committee has been observing your work and has confidence in your abilities and potential. We would like to give you an opportunity for advancement. Cultures that live in constant threat of natural disaster have certain mechanisms in place that let them live productively and even let them return to a site after a disaster, in the face of recurrence.

- Please investigate the factors and forces that motivate a civilization to live this way.

- Try to find patterns of behavior and report them to us.

Specific Physical Geography

Your Assignment: The Committee is planning a fact-finding trip with certain world leaders to learn about the use of rivers in all facets of daily life. Your job is to visit the following rivers at their mouths, follow them to their sources, and report how accessible and navigable they are. Be prepared to discuss how they are used and who uses them. Find as many different uses as you can. Follow all orders and good luck.

0°10'S	49°00'W	52°56'N	141°10'E
31°48'N	121°10'E	6°04'S	12°24'E
45°30'N	29°45'E	72°25'N	126°40'E
10°15'N	105°55'E	29°10'N	89°15'W
38°50'N	90°08'W	31°20'N	31°00'E
66°45'N	69°30'E	33°43'S	59°15'W

Special Instructions: The Committee has been observing your work and has confidence in your abilities and potential. We would like to give you an opportunity for advancement. The people who live by a great river have a lot in common with people who live by the ocean or a great lake; their way of life is tied to the water.

- Please find out about the ways of life on two of the rivers that interest you.

- Compare and contrast them and tell us what you believe to be the essential nature of life on a river.

Gumshoe Geography © 1996 Zephyr Press, Tucson, AZ

Specific Physical Geography

Your Assignment: The Committee is interested in studying the power of spiritual belief on a population. Your job is to visit the following sacred mountains and find out about the geography of the peak and the region around each site. Be prepared to discuss the common physical characteristics among all of the sites. Follow all orders and good luck.

32°39'S	70°00'W	39°40'N	44°24'E
54°10'N	449°40'E	3°04'S	37°22'E
13°07'S	72°34'W	45°58'N	7°39'E
32°25'S	116°18'E	35°26'N	138°43'E
45°55'N	68°55'W		

Special Instructions: The Committee has been observing your work and has confidence in your abilities and potential. We would like to give you an opportunity for advancement. For centuries, people have invented stories and legends about aspects of their environment that hold them in awe.

- Please collect some of the stories and legends from the sites that you visit.

- Let us know which cultures are still in existence and how the people feel about the mountains. In what ways have their attitudes been changed by Western influences?

Specific Physical Geography

Your Assignment: Your job is to visit the following sacred mountains and find out about the geography of the peak and the region around each site. Be prepared to tell us what makes the mountain in question different from the other mountains in the area or from the surrounding terrain. Follow all orders and good luck.

63°30′N	151°00′W	40°05′N	22°21′E
41°20′N	122°20′W	28°32′N	33°59′E
31°46′N	35°14′E	38°51′N	105°03′W
19°02′N	98°38′W	7°58′S	113°35′E

Special Instructions: The Committee has been observing your work and has confidence in your abilities and potential. We would like to give you an opportunity for advancement.

- Please write a profile about the culture that believed or believes in the mountain as a sacred mountain.
- Let us know if the mountain stood alone as an icon or if it was part of a system of belief.
- Let us know how the belief system worked.

Gumshoe Geography © 1996 Zephyr Press, Tucson, AZ

Specific Physical Geography

Your Assignment: The Committee is in the process of establishing monitoring installations near certain nations. Your job is to visit the following offshore islands and find out about the ways in which the islands are managed and administered. Be prepared to discuss the differences between the ones that are independent and the ones that are controlled by continental governments. Let us know which ones would be friendly and which ones wouldn't. Follow all orders and good luck.

4°30'N	9°30'E	42°30'S	73°55'W
19°00'N	109°00'E	54°15'N	4°30'W
40°50'N	73°00'W	36°10'N	28°00'E
33°30'N	133°30'E	37°30'N	14°00'E
49°16'N	123°07'W	6°10'S	39°20'E

Special Instructions: The Committee has been observing your work and has confidence in your abilities and potential. We would like to give you an opportunity for advancement. It is often the case that even though an island is close to the mainland, it has a separate and unique culture.

- Please find out how true this statement is about the sites that you are visiting.
- Find out why the differences or similarities exist and what causes them.

Specific Physical Geography

Your Assignment: The Committee is investigating the option of harnessing the power of natural waterfalls to add to the world's power grid. Your job is to visit the following sites and find out about the geography of the sites. Be prepared to discuss how navigable the approach is to the sites and how high and wide the waterfall is. Find out what caused the formation of the waterfall. Follow all orders and good luck.

5°57′N	62°30′W	47°13′N	12°11′E
44°48′S	167°44′E	55°30′N	126°00′W
29°14′S	31°30′E	35°28′N	119°33′W

Following is a list of other waterfalls for which you might look.

Cascade de Giétroz: Switzerland
Cuguenán: Venezuela
East Mardalsfoss: Norway
Gavarnie: France
Great Falls: Guyana
Ribbon Falls: California

Special Instructions: The Committee has been observing your work and has confidence in your abilities and potential. We would like to give you an opportunity for advancement.

- Please research the history of each site and find out about the exploration and discovery of each waterfall.

- Also please find out about the local population today, how they live, and how the waterfall is part of their lives.

Gumshoe Geography © 1996 Zephyr Press, Tucson, AZ

Specific Physical Geography

Your Assignment: The Committee is helping a certain group of nations design a model for dealing with flooding and river drainage at delta regions. Your job is to visit the following sites and find out about how the region changes, how often it floods, and what efforts are being made to change the course of the river to serve other purposes. Be prepared to cite environmentally sound examples of dealing with a delta region. Follow all orders and good luck.

0°10′S	49°00′W	15°50′N	95°06′E
45°30′N	29°45′E	23°20′N	90°30′E
69°00′N	136°30′W	10°20′N	106°40′E
29°10′N	89°15′W	14°30′N	4°00′W
4°50′N	6°00′E	31°20′N	31°00′E
8°37′N	62°15′W	51°52′N	6°02′E

Special Instructions: The Committee has been observing your work and has confidence in your abilities and potential. We would like to give you an opportunity for advancement. Delta regions are often flooded and unpredictable.

- Please find out about the people who live in the regions on the list and chart the different ways of dealing with this unique habitat.

- Also find out about the animal life that thrives in these sites.

Specific Physical Geography

Your Assignment: The Committee has been interested in the use of deep ocean trenches for some time now. It appears that the time is right for closer investigation. Your job is to investigate the following sites and find out the extent of exploration of each site. Be prepared to detail the types of animal and plant life that thrive in each site and to make comparisons among the sites. Follow all orders and good luck.

30°00′N	145°00′E	37°00′N	143°00′E
30°00′S	177°00′W	47°00′N	150°00′E
14°00′N	147°30′E	6°00′S	153°00′E
9°00′N	127°00′E	21°00′S	175°00′W
8°30′N	138°00′E		

Special Instructions: The Committee has been observing your work and has confidence in your abilities and potential. We would like to give you an opportunity for advancement.

- Please find out what it takes to explore these types of environments.
- Find out when exploration started and how the technology has improved over the years.
- Find out who the major players have been and find out a little about their lives.

   Gumshoe Geography © 1996 Zephyr Press, Tucson, AZ

Specific Physical Geography

Your Assignment: The Committee has been interested in the use of deep ocean trenches for some time now. It appears that the time is right for closer investigation. Your job is to visit the following sites and determine if the proximity to an ocean trench causes any noticeable changes in the ocean surface or the landforms that are nearby. Be prepared to discuss the geological mechanisms that cause ocean trenches. Follow all orders and good luck.

51°00'N	179°00'E	19°00'N	80°00'W
33°00'S	90°00'W	10°30'S	110°00'E
15°00'N	95°00'W	20°00'S	168°00'E
6°30'N	134°30'E	20°00'N	66°00'W
25°45'N	128°00'E	56°30'S	25°00'W

Special Instructions: The Committee has been observing your work and has confidence in your abilities and potential. We would like to give you an opportunity for advancement. In a world of high population growth, waste and trash disposal become a more and more pressing problem.

- Please investigate the use of ocean trenches for this purpose.
- Find out about any present usage and categorize the positive and negative effects.
- Please prepare to debate one side or the other.

9

HUMAN-MADE GEOGRAPHY

World Map Activity

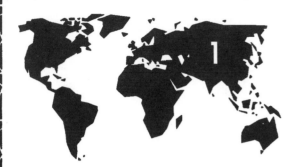

Human-Made Geography

Your Assignment: The Committee is preparing to make a case before the World Court against the continued nationalistic use of fresh water resources in the world. Your job is to visit the following dam sites and find out who manages and administers the dams. Find out the volumes they hold, their ages, types of construction, and any other relevant information. Be prepared to argue for either side of the question of whether or not they should be administered by an international organization. Follow all orders and good luck.

31°00′N	47°25′E	56°14′N	117°17′W
15°43′S	32°42′E	51°30′N	68°19′W
47°45′N	106°50′W	24°01′N	32°52′E
36°00′N	114°27′W		

Special Instructions: The Committee has been observing your work and has confidence in your abilities and potential. We would like to give you an opportunity for advancement. One of the outcomes of building a dam is a lake.

- Please tell us about the lakes that are associated with the dams on the list and how the lakes are used.

- Tell us about recreational uses, and if you can find stories about the taking of land to build a dam and reservoir, please let us know about them.

Human-Made Geography

Your Assignment: The Committee is involved in the international shipping trade and we would like to investigate the viability of constructing canals at strategic locations. Your job is to visit the following sites and report on the history of the construction of these canals. Be prepared to talk about the reasons for building them, the problems of constructing them, and the demands of administering them. Follow all orders and good luck.

50°39'N	5°37'E	51°57'N	5°25'E
30°05'N	94°06'W	43°14'N	79°13'W
54°20'N	10°08'E	9°20'N	79°55'W
49°15'N	67°00'W	29°55'N	32°33'E

Special Instructions: The Committee has been observing your work and has confidence in your abilities and potential. We would like to give you an opportunity for advancement. Working canals and locks is a process that requires many specialized jobs.

- Please draw a graphic explanation of how locks work and profile the people who work them.
- Contrast canal work of the past with the job requirements at present.

Human-Made Geography

Your Assignment: The Committee is working with a multinational corporation that is seeking to gain control of many of the strategic rail tunnels in the world. Your job is to visit the following sites and find out how they were built and what technology was used. Be prepared to discuss the improvements in technology and how technology has made the construction of tunnels easier and safer. Follow all orders and good luck.

44°29′N	11°20′E	47°08′N	10°12′E
55°20′N	3°27′W	45°15′N	6°54′E
44°31′N	115°59′W	41°15′S	175°15′E
41°40′N	140°55′E	46°15′N	8°00′E
46°30′N	8°30′E	48°30′N	7°10′E

Special Instructions: The Committee has been observing your work and has confidence in your abilities and potential. We would like to give you an opportunity for advancement. Before the construction of the tunnel at each site, transportation needs had to be met in a different way.

- Please let us know how goods and people were transported before the construction of each tunnel.

- Let us know how much mileage and time is saved by the tunnel.

Human-Made Geography

Your Assignment: The Committee is in the process of doing a feasibility study for a new tunnel for vehicles between two sections of a certain nation. Your job is to visit the following sites and find out how the tunnels were constructed in general. Be prepared to tell us how the tunnels were financed, who controlled the construction, and who now controls the tunnel. Follow all orders and good luck.

39°17′N	76°37′W	45°50′N	8°10′E
37°02′N	76°23′W	40°43′N	74°01′W
45°50′N	6°52′E	45°31′N	73°34′W
53°25′N	2°55′W	46°30′N	8°30′E

Special Instructions: The Committee has been observing your work and has confidence in your abilities and potential. We would like to give you an opportunity for advancement. The opening of a vehicular tunnel anywhere in the world usually means the opening of certain new commercial markets and opportunities.

- Please investigate any new recreational, commercial, or personal opportunities that might have opened with the opening of these tunnels.

World Map Activity

Human-Made Geography

Your Assignment: The Committee has a strong interest in certain archaeological finds that have appeared in the world. Your job is to visit the following sites and find out how much of the original site is still recognizable and available for study. Be prepared to discuss efforts, both past and present, to uncover more of the site for scientific study. Follow all orders and good luck.

13°26′N	103°52′E	32°32′N	44°25′E
20°40′N	88°35′W	35°18′N	25°10′W
15°30′N	45°21′E	13°07′S	72°34′W
6°55′N	158°15′E	30°03′N	31°15′E
27°07′S	109°22′W	51°11′N	1°49′W
17°20′N	89°39′W	50°41′N	4°46′W
39°57′N	26°15′E	20°16′S	30°55′E

Special Instructions: The Committee has been observing your work and has confidence in your abilities and potential. We would like to give you an opportunity for advancement. The sites that you are to visit were once thriving centers of civilizations.

- Please find out about them and about all facets of the culture from family life to government.

- Write a fictional account of a visit to the site during its golden age.

Human-Made Geography

Your Assignment: The Committee is helping a certain government design a model for a new type of airport that would help relieve the congestion at the airports on your list. Your job is to visit the following sites, report on the layout of each site, and list them by volume or air traffic. Tell us about the considerations that went into designing each facility and how well it works. Be prepared to discuss the good and bad points of each facility and why these cities would become hubs of air travel. Propose your own ideas for a state-of-the-art facility. Follow all orders and good luck.

33°45′N	82°43′W	48°52′N	2°20′E
32°47′N	96°48′W	42°45′N	12°29′E
52°21′N	14°33′E	51°30′N	0°10′W
35°40′N	139°46′E	51°30′N	0°10′W
22°15′N	114°10′E	34°03′N	118°15′W
43°39′N	79°23′W	35°40′N	139°46′E
41°53′N	87°38′W	48°52′N	2°20′E
35°37′N	137°14′E		

Special Instructions: The Committee has been observing your work and has confidence in your abilities and potential. We would like to give you an opportunity for advancement. One of the toughest jobs in the world is that of an air traffic controller.

- Please find out how air traffic controllers are trained and what they go through on a normal day on the job.

- Find out what makes their job more difficult and what makes it easier.

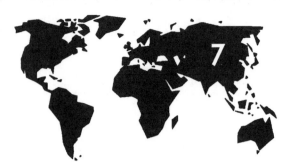

Human-Made Geography

Your Assignment: The Committee is publishing a catalogue of bridge construction for certain nations to help them upgrade their bridges' infrastructure. Your job is to visit the following bridges and find out what type of bridge each is, what it was designed for, how old it is, how it is holding up, and how it fits its surroundings. Be prepared to compare and contrast bridge construction methods and cite the good and bad points of each one. Follow all orders and good luck.

49°16'N	123°07'W	41°09'N	8°37'W
46°11'N	123°50'W	41°00'N	29°00'E
13°32'N	100°36'E	37°49'N	122°29'W
22°35'N	88°20'E	53°40'N	0°10'W
33°56'N	130°57'E	45°48'N	16°00'E
32°48'N	79°57'W	35°37'N	137°14'E
46°49'N	71°13'W	68°26'N	17°25'E
33°52'S	151°13'E	9°20'N	79°55'W
58°00'N	11°38'E	40°43'N	74°01'W

Special Instructions: The Committee has been observing your work and has confidence in your abilities and potential. We would like to give you an opportunity for advancement. A bridge is always a wonder to behold, both during construction and when it is finished.

- Please report on the history of the construction of each bridge.
- Tell us about the reasons it was constructed, the people who built it, the problems they encountered, and the technology that they used.
- Please try to find information on great bridge architects, also.

Human-Made Geography

Your Assignment: The Committee has always been interested in creating something that will live in perpetuity. Your job is to visit the following sites and find out why they are included in the list of the Wonders of the World. Be prepared to discuss the factors that set them apart and to tell us about their construction, their longevity, and about any remaining evidence of their past glory. Follow all orders and good luck.

36°10′N	28°00′E	32°32′N	44°25′E
37°02′N	27°06′E	31°12′N	29°54′E
40°05′N	22°21′E	37°55′N	27°20′E
30°03′N	31°15′E		

Special Instructions: The Committee has been observing your work and has confidence in your abilities and potential. We would like to give you an opportunity for advancement. Building places such as these took a great deal of foresight and effort on the part of the local population.

- Please investigate and report on the people, both those in charge and those who did the actual construction, who built these monuments.

- Speculate as to why these monuments were built and how they affected the culture of the region that surrounded them.

10

HISTORICAL PERIODS AND EVENTS

World Map Activity

Historical Periods and Events

Your Assignment: The Committee is testing a new method of historical exploration. You will be sent back to the year 1880 and will visit the following locations. You will determine from your observations what factors and characteristics made these places the large metropolitan centers that they are now. Follow all instructions and good luck.

39°06′N	84°31′W	39°05′N	94°35′W
35°08′N	90°03′W	45°33′N	122°36′W
38°35′N	121°30′W	38°38′N	90°11′W
40°46′N	111°53′W	29°28′N	98°31′W
35°41′N	105°56′W	37°41′N	97°20′W

Special Instructions: The Committee has been observing your work and has confidence in your abilities and potential. We would like to give you an opportunity for advancement.

- Please draw a detailed map of the area that you explored in traveling among your destinations. Make the map as it would have been made in that day, with pictures of hazards and obstacles, rivers and mountains, and other points of interest along the route.

Gumshoe Geography © 1996 Zephyr Press, Tucson, AZ

Historical Periods and Events

Your Assignment: The Committee is interested in knowing about the evolution of farming methodology in order to help us project the future trends in that industry. Your job is to visit the following sites and report on the methods and crops that were characteristic of the sites. Be prepared to discuss early tools and the hardships cultivation presented in these places. Follow all instructions and good luck.

20°N	100°W	30°N	30°E
39°N	22°E	45°N	35°E
40°N	110°E	30°N	70°E
5°S	75°W	15°N	100°E
38°N	35°W	10°N	0°

Special Instructions: The Committee has been observing your work and has confidence in your abilities and potential. We would like to give you an opportunity for advancement.

- Please investigate the cultures of these sites and how they were affected by the change to a farming society from that of hunter-gatherer.
- Let us know how the change affected family life and the development of civilization at those sites.

Historical Periods and Events

Your Assignment: The Committee is looking for common ground among a group of nations that are not presently at peace. Your job is to visit the following sites and report on the common Roman roots of each site. Be prepared to discuss the existing evidence of Roman culture, both architectural and cultural. Follow all instructions and good luck.

31°12'N	20°54'E	36°14'N	36°07'E
37°59'N	23°44'E	47°30'N	19°05'E
36°49'N	10°18'E	32°48'N	21°59'E
31°46'N	35°14'E	38°43'N	9°08'W
51°30'N	0°10'W	45°45'N	4°51'E
43°18'N	5°24'E	48°52'N	2°20'E
42°45'N	12°29'E	37°04'N	15°18'E
39°50'N	4°00'W	48°12'N	16°22'E

Special Instructions: The Committee has been observing your work and has confidence in your abilities and potential. We would like to give you an opportunity for advancement.

- Please let us know what it was like to be a Roman citizen in any of the sites.

- Please try to find out if there was something essential to being a Roman, no matter where you were in the empire.

Historical Periods and Events

Your Assignment: The Committee is studying the economics of a certain region and we could use the data that you are being sent to collect. Your job is to visit the following sites and report on the types of goods and services that were traded in this region during the early Middle Ages. Be prepared to discuss the modes of exchange and who the traders were at the time. Follow all instructions and good luck.

31°12′N	20°54′E	51°13′N	3°14′E
41°01′N	28°58′E	44°25′N	8°57′E
53°33′N	10°00′E	50°26′N	30°31′E
51°30′N	0°10′W	43°18′N	5°24′E
40°50′N	14°15′E	58°31′N	31°17′E
48°52′N	2°20′E	43°43′N	10°23′E
42°45′N	12°29′E	59°20′N	18°03′E
45°27′N	12°21′E		

Special Instructions: The Committee has been observing your work and has confidence in your abilities and potential. We would like to give you an opportunity for advancement.

- Please write a fictionalized personal account of a trading trip at this time. Include information on the modes of transporting the goods over the sea and land trade routes.

Historical Periods and Events

Your Assignment: The Committee has been investigating the origin of certain rare artifacts. Your job is to visit the following sites and report on the role that each played in the Crusades. Be prepared to discuss the reasons for the Crusades, the planned objectives, and how well they were carried out. Follow all instructions and good luck.

36°14′N	36°07′E	41°01′N	28°58′E
44°25′N	8°57′E	31°46′N	35°14′E
45°45′N	4°51′E	43°18′N	5°24′E
49°08′N	6°10′E	48°52′N	2°20′E
43°43′N	10°23′E	42°45′N	12°29′E
34°26′N	35°51′E	45°27′N	12°21′E
48°12′N	16°22′E	39°55′N	37°48′E

Special Instructions: The Committee has been observing your work and has confidence in your abilities and potential. We would like to give you an opportunity for advancement.

- Please investigate the people who went on the Crusades.
- Find out who their leaders were and what they were thinking.
- Find out what it was like to live in that day and what would motivate a person to go on such a journey.

Gumshoe Geography © 1996 Zephyr Press, Tucson, AZ

Historical Periods and Events

Your Assignment: The Committee is looking into the ancient heritages of certain European peoples to try to help solve particular impasses in current negotiations. Your job is to visit some well-known sites in Europe, find out about the barbarian tribes that are listed below, and find out how they ruled the land after the fall of the Roman Empire. Be prepared to show their territory and to discuss who they were and how they lived. Follow all instructions and good luck.

Anglo-Saxons	Basques
Burgundians	Byzantines
Franks	Huns
Jutes	Magyars
Muslims	Ostrogoths
Picts	Scots
Slavs	Vandals
Vikings	Visigoths

Special Instructions: The Committee has been observing your work and has confidence in your abilities and potential. We would like to give you an opportunity for advancement.

- Please find out how these people lived and if they really deserved the reputation which the modern usage of the word *barbarian* suggests.

- Find out who their leaders were and what they wanted for their people.

Historical Periods and Events

Your Assignment: The Committee is investigating the British Empire at its peak to find out which of their strategies were successful and which were not. Your job is to visit the following sites and report on the reasons that the British wanted to control each site. Be prepared to discuss the resources that were valuable to the British and how they made use of those resources. Follow all instructions and good luck.

25°00'S	135°00'E	17°35'N	88°35'W
22°00'S	24°00'E	60°00'N	95°00'W
12°00'N	15°00'W	5°00'N	59°00'W
13°30'S	34°00'E	41°00'S	174°00'E
10°00'N	8°00'E	30°00'S	26°00'E
8°30'N	11°30'W	8°00'N	1°10'E
15°00'S	30°00'E	20°00'S	30°00'E

Special Instructions: The Committee has been observing your work and has confidence in your abilities and potential. We would like to give you an opportunity for advancement.

- Please investigate how the British maintained their possessions in the empire.
- Find out how they governed them and what strategies were necessary to keep them as colonies.
- Please cite the reasons for the eventual downfall of the empire.

Gumshoe Geography © 1996 Zephyr Press, Tucson, AZ

Historical Periods and Events

Your Assignment: The Committee is doing some work for a certain monarchy and needs the information that you are being sent to collect. Your job is to visit the following sites and report on their acquisition and inclusion into the British Empire. Be prepared to discuss the people in power at that time and what their reasoning and purpose was for imperialism. Follow all instructions and good luck.

22°00'N	98°00'E	27°00'N	30°00'E
22°00'N	77°00'E	1°00'N	38°00'E
4°00'N	102°00'E	21°00'N	57°00'E
30°00'N	70°00'E	6°00'S	150°00'E
10°00'N	49°00'E	15°00'N	30°00'E
43°00'S	147°00'E	1°02'N	32°00'E
15°00'N	44°00'E		

Special Instructions: The Committee has been observing your work and has confidence in your abilities and potential. We would like to give you an opportunity for advancement.

- Please investigate the continued British influence on the way of life of the native residents of the former British colonies.

- Please let us know in which places the British influence is stronger and give us your opinion as to why this is so.

Historical Periods and Events

Your Assignment: The Committee is helping a certain private firm locate and identify early sites of exploration in the Western Hemisphere. Your job is to visit the following sites and report on who explored them and where the explorers were from. Be prepared to discuss each expedition and how it fared. Follow all instructions and good luck.

6°15′S	78°50′W	46°00′N	60°30′E
13°31′S	71°59′W	42°20′N	83°03′W
45°00′N	87°30′W	45°31′N	73°34′W
29°58′N	90°07′W	9°00′N	80°00′W
46°49′N	71°13′W	32°43′N	117°09′W
13°42′N	89°12′W	29°51′N	81°25′W

Special Instructions: The Committee has been observing your work and has confidence in your abilities and potential. We would like to give you an opportunity for advancement.

- Please investigate the purpose of each exploration in the minds of its leaders, and find out what the interactions between the explorers and the natives were like.

- Find out which groups got along with the native people and why, as well as which groups did not and why not.

Gumshoe Geography © 1996 Zephyr Press, Tucson, AZ

Historical Periods and Events

Your Assignment: The Committee is helping to negotiate a settlement in a conflict that has so far resisted solution. We need the background information that you will find to complete our study. Your job is to visit the following sites and report on the importance of each site in what was called the Great War. Be prepared to discuss the reasons for the conflict and how it changed the way of life on the continent of Europe. Follow all instructions and good luck.

49°54′N	2°18′E	54°05′N	8°50′E
49°03′N	3°24′E	40°03′N	17°58′E
53°57′N	9°00′E	53°45′N	21°45′E
48°54′N	5°33′E	43°50′N	18°25′E
49°55′N	2°30′E	59°55′N	30°15′E
49°10′N	5°23′E		

Special Instructions: The Committee has been observing your work and has confidence in your abilities and potential. We would like to give you an opportunity for advancement.

- Please investigate the innovations in the way that war was fought during this period.

- Tell us who invented and used the new methods and tools of war and how successful they were.

Historical Periods and Events

Your Assignment: The Committee is working on a catalogue of the Pacific Ocean and its islands. Your job is to visit the following sites and report on the history of the islands with respect to their part in World War II. Be prepared to discuss the battles that took place at these sites, how long they lasted, what the strategies were, and what the costs were. Follow all instructions and good luck.

20°30′N	121°50′E	20°00′S	158°00′E
14°35′N	121°00′E	9°32′S	160°12′E
13°28′N	144°47′E	34°24′N	132°27′E
24°47′N	141°20′E	10°50′N	124°50′E
28°13′N	177°22′W	32°47′N	129°56′E
26°20′N	127°47′E	21°20′N	158°00′W

Special Instructions: The Committee has been observing your work and has confidence in your abilities and potential. We would like to give you an opportunity for advancement.

- Please investigate the war effort in the United States and how the American people created a culture of war around the conflict in the Pacific.

- Find examples of attempts to glamorize the war and also examples of realistic portrayal of the war.

Gumshoe Geography © 1996 Zephyr Press, Tucson, AZ

Historical Periods and Events

Your Assignment: The Committee is helping certain markets in Europe overcome obstacles. Your job is to visit the following sites and report on their importance in World War I and how they played a part in the Axis and Allied plans. Be prepared to discuss the expected final outcome of each plan and how the people in charge worked to bring it about. Follow all instructions and good luck.

48°54′N	5°33′E	43°50′N	18°25′E
49°55′N	2°30′E	59°55′N	30°15′E
49°10′N	5°23′E	42°29′N	79°21′W
30°99′N	28°57′E	59°55′N	30°15′E
51°30′N	0°10′W	36°00′N	0°35′W
48°52′N	2°20′E	50°34′N	7°14′E
42°45′N	12°29′E	48°44′N	44°25′E
32°05′N	23°59′E	34°00′N	9°00′E

Special Instructions: The Committee has been observing your work and has confidence in your abilities and potential. We would like to give you an opportunity for advancement.

- Please investigate the war strategy of each side and create a time line showing the play and counterplay nature of the war.

- Highlight strategic mistakes and examples of good fortune for both sides.

11
WORLD
EXPLORATION

World Map Activity

World Map Activity

World Exploration

Henry Hudson

Your Assignment: The Committee has decided that the explorations of Henry Hudson have a bearing on the success of our mission. Your job is to find and visit six sites that he may have visited. Your report should include one or all of the following:

1. Find six or more sites and explain their importance to his explorations.
2. Draw an annotated map of the journey or journeys; include a key.
3. Report on the results of each journey and how it changed the course of history.
4. Report on any new plant or animal species discovered during the journeys.
5. Report on the hardships that the explorers encountered.

Special Instructions: The Committee has been observing your work and has confidence in your abilities and potential. We would like to give you an opportunity for advancement.

- Please investigate the exploration in more detail and include any or all of the following in your report:

1. Draw and describe in detail the modes of transportation and any other examples of technology that were used on the journeys.
2. Report on the political climate in the world and in the area, and on any other significant world events that were happening at that time.
3. Write a fictional account of any part of the exploration that interests you.
4. Write a personality profile of the major players in the exploration and describe the character of the journeys, including explorers' attitudes toward the cultures of the native peoples.

Gumshoe Geography © 1996 Zephyr Press, Tucson, AZ

World Map Activity

World Exploration

Ibn Battuta

Your Assignment: The Committee has decided that the explorations of Ibn Battuta have a bearing on the success of our mission. Your job is to visit the following sites and report on them. Your report should include one or more of the following:

1. Find the sites listed and explain their importance to the exploration.
2. Draw an annotated map of the journey or journeys; include a key.
3. Report on the results of each journey and how it changed the course of history.
4. Report on any new plant or animal species discovered during the journeys.
5. Report on the cultures that the explorers encountered during their exploration.

39°00'N	35°00'E	23°07'N	113°18'E
42°30'N	45°00'E	45°00'N	34°00'E
22°00'N	77°00'E	21°29'N	39°12'E
9°17'S	28°20'E	21°27'N	39°49'E
4°03'S	39°40'E		

Special Instructions: The Committee has been observing your work and has confidence in your abilities and potential. We would like to give you an opportunity for advancement.

- Please investigate the exploration in more depth and include any or all of the following in your report:

1. Draw and describe in detail the modes of transportation and any other examples of technology that were used on the journeys.
2. Report on the political climate in the world and in the area and on any other significant world events that were happening at that time.
3. Write a fictional account of any part of the exploration that interests you.
4. Write a personality profile of the major players in the exploration and describe the character of the journey, including the explorers' attitude toward the cultures of the native peoples.

World Exploration

Richard Burton and John Speke

Your Assignment: The Committee has decided that the explorations of Burton and Speke have a bearing on our mission. Your job is to visit the following sites and report on them. Your report should include one or all of the following:

1. Find the sites listed and explain their importance to the exploration.
2. Draw an annotated map of the journey or journeys; include a key.
3. Report on the results of each journey and how it changed the course of history.
4. Report on any new plant or animal species discovered during the journeys.
5. Report on the cultures that the explorers encountered during their exploration.

2°30′S	32°54′E	1°01′N	32°57′E
5°20′S	32°30′E	4°55′S	29°41′E
6°10′S	39°20′E		

Special Instructions: The Committee has been observing your work and has confidence in your abilities and potential. We would like to give you an opportunity for advancement.

■ Please investigate the exploration in more detail and include any or all of the following in your report:

1. Draw and describe in detail the modes of transportation and any other examples of technology that were used on the journeys.
2. Report on the political climate in the world and in the areas and on any other significant world events that were happening at that time.
3. Write a fictional account of any part of the exploration that interests you.
4. Write a personality profile of the major players in the exploration and describe the character of the journeys, including the explorers' attitude toward the cultures of the native peoples.

Gumshoe Geography © 1996 Zephyr Press, Tucson, AZ

World Exploration

Henry Stanley and David Livingstone

Your Assignment: The Committee has decided that the explorations of Stanley and Livingstone have a bearing on our mission. Your job is to visit the following sites and report on them. Your report should include one or more of the following:

1. Find the sites listed and explain their importance to the exploration.
2. Draw an annotated map of the journey or journeys; include a key.
3. Report on the results of each journey and how it changed the course of history.
4. Report on any new plant or animal species discovered during the journeys.
5. Report on the cultures that the explorers encountered during their exploration.

5°51'S	13°03'E	12°00'S	34°30'E
20°37'S	22°40'E	12°19'S	30°18'E
0°26'N	25°20'E	5°20'S	32°30'E
4°55'S	29°41'E	17°56'S	25°50'E
6°04'S	12°24'E	6°10'S	39°20'E

Special Instructions: The Committee has been observing your work and has confidence in your abilities and potential. We would like to give you an opportunity for advancement.

- Please investigate the exploration in more depth and include any or all of the following in your report:

1. Draw and describe in detail the modes of transportation and any other examples of technology that were used on the journeys.
2. Report on the political climate in the world and in the areas and on any other significant world events that were happening at that time.
3. Write a fictional account of any part of the exploration that interests you.
4. Write a personality profile of the major players in the exploration and describe the character of the journey, including explorer's attitudes toward the cultures of the native peoples.

World Exploration

Hsuan-tsang

Your Assignment: The Committee has decided that the explorations of Hsuan-tsang have a bearing on the success of our mission. Your job is to find and visit six sites that he may have visited. Your report should include one or more of the following:

1. Find the sites listed and explain their importance to the exploration.
2. Draw an annotated map of the journey or journeys; include a key.
3. Report on the results of each journey and how it changed the course of history.
4. Report on any new plant or animal species discovered during the journeys.
5. Report on the cultures that the explorers encountered during their exploration.

25°27′N	81°51′E	25°00′N	93°00′E
39°55′N	116°23′W	43°00′N	106°00′E
35°00′N	71°00′E	24°20′N	67°47′E
34°00′N	76°00′E		

Special Instructions: The Committee has been observing your work and has confidence in your abilities and potential. We would like to give you an opportunity for advancement.

■ Please investigate the exploration in more detail and include any or all of the following in your report:

1. Draw and describe in detail the modes of transportation and any other examples of technology that were used on the journeys.
2. Report on the political climate in the world and in the area and on any other significant world events that were happening at that time.
3. Write a fictional account of any part of the exploration that interests you.
4. Write a personality profile of the major players in the exploration and describe the character of the journey, including explorers' attitudes toward the cultures of the native peoples.

World Exploration

Alexine Tinne

Your Assignment: The Committee has decided that the explorations of Alexine Tinne have a bearing on the success of our mission. Your job is to visit the following sites and report on them. Your report should include one or more of the following:

1. Find the sites listed and explain their importance to the exploration.
2. Draw an annotated map of the journey or journeys; include a key.
3. Report on the results of each journey and how it changed the course of history.
4. Report on any new plant or animal species discovered during the journeys.
5. Report on the cultures that the explorers encountered during their exploration.

27°10′N	31°16′E	30°03′N	31°15′E
24°05′N	32°53′E	15°50′N	33°00′E
52°06′N	4°18′E		

Special Instructions: The Committee has been observing your work and has confidence in your abilities and potential. We would like to give you an opportunity for advancement.

- Please investigate the exploration in more depth and include any or all of the following in your report:

  1. Draw and describe in detail the modes of transportation and any other examples of technology that were used on the journeys.
  2. Report on the political climate in the world and in the areas and on any other significant world events that were happening at that time.
  3. Write a fictional account of any part of the exploration that interests you.
  4. Write a personality profile of the major players in the exploration and describe the character of the journey, including explorers' attitudes toward the cultures of the native peoples.

World Exploration

Mary Kingsley

Your Assignment: The Committee has decided that the explorations of Mary Kingsley have a bearing on the success of our mission. Your job is to visit the following sites and report on them. Your report should include one or more of the following:

1. Find the sites listed and explain their importance to the exploration.
2. Draw an annotated map of the journey or journeys; include a key.
3. Report on the results of each journey and how it changed the course of history.
4. Report on any new plant or animal species discovered during the journeys.
5. Report on the cultures that the explorers encountered during their exploration.

28°00′N	15°30′W
4°12′N	9°11′E
0°49′S	9°00′E

Special Instructions: The Committee has been observing your work and has confidence in your abilities and potential. We would like to give you an opportunity for advancement.

- Please investigate the exploration in more depth and include any or all of the following in your report:

1. Draw and describe in detail the modes of transportation and any other examples of technology that were used on the journeys.
2. Report on the political climate in the world and in the areas and on any other significant world events that were happening at that time.
3. Write a fictional account of any part of the exploration that interests you.
4. Write a personality profile of the major players in the exploration and describe the character of the journey, including the explorers' attitudes toward the cultures of the native peoples.

World Exploration

May French Sheldon

Your Assignment: The Committee has decided that the explorations of May French Sheldon have a bearing on our mission. Your job is to visit the following sites and report on them. Your report should include one or more of the following:

1. Find the sites listed and explain their importance to the exploration.
2. Draw an annotated map of the journey or journeys; include a key.
3. Report on the results of each journey and how it changed the course of history.
4. Report on any new plant or animal species discovered during the journeys.
5. Report on the cultures that the explorers encountered during their exploration.

7°42′S	30°00′E	4°03′S	39°40′E
3°04′S	37°22′E	5°26′S	38°58′E
3°24′S	37°41′E	6°10′S	39°20′E

Special Instructions: The Committee has been observing your work and has confidence in your abilities and potential. We would like to give you an opportunity for advancement.

- Please investigate the exploration in more depth and include any or all of the following in your report:

1. Draw and describe in detail the modes of transportation and any other examples of technology that were used on the journeys.
2. Report on the political climate in the world and in the areas and on any other significant world events that were happening at that time.
3. Write a fictional account of any part of the exploration that interests you.
4. Write a personality profile of the major players in the exploration and describe the character of the journey, including the explorers' attitudes toward the cultures of the native peoples.

World Exploration
Delia Denning Akeley

Your Assignment: The Committee has decided that the explorations of Delia Denning Akeley have a bearing on our mission. Your job is to visit the following sites and report on them. Your report should include one or more of the following:

1. Find the sites listed and explain their importance to the exploration.
2. Draw an annotated map of the journey or journeys; include a key.
3. Report on the results of each journey and how it changed the course of history.
4. Report on any new plant or animal species discovered during the journeys.
5. Report on the cultures that the explorers encountered during their exploration.

Beaver Dam, Wisconsin	Ituri Forest, Zaire
Mt. Meru, Kenya	Muddo Gashi
Pygmys	San Kuri
Tama River, Kenya	

Special Instructions: The Committee has been observing your work and has confidence in your abilities and potential. We would like to give you an opportunity for advancement.

- Please investigate the exploration in more depth and include any or all of the following in your report:

1. Draw and describe in detail the modes of transportation and any other examples of technology that were used on the journeys.
2. Report on the political climate in the world and in the areas and on any other significant world events that were happening at that time.
3. Write a fictional account of any part of the exploration that interests you.
4. Write a personality profile of the major players in the exploration and describe the character of the journey, including the explorers' attitudes toward the cultures of the native peoples.

     Gumshoe Geography © 1996 Zephyr Press, Tucson, AZ

World Exploration

Dame Freya Madeline Stark

Your Assignment: The Committee has decided that the explorations of Dame Freya Madeline Stark have a bearing on our mission. Your job is to visit the following sites and report on them. Your report should include one or all of the following:

1. Find the sites listed and explain their importance to the exploration.
2. Draw an annotated map of the journey or journeys; include a key.
3. Report on the results of each journey and how it changed the course of history.
4. Report on any new plant or animal species discovered during the journeys.
5. Report on the cultures that the explorers encountered during their exploration.

12°46′N	45°01′E	33°21′N	44°23′E
30°03′N	31°15′E	44°28′N	7°22′E
44°39′N	63°36′W	33°30′N	48°40′E
36°00′N	54°00′E	48°52′N	2°20′E

Special Instructions: The Committee has been observing your work and has confidence in your abilities and potential. We would like to give you an opportunity for advancement.

- Please investigate the exploration in more depth and include any or all of the following in your report:

  1. Draw and describe in detail the modes of transportation and any other examples of technology that were used on the journeys.
  2. Report on the political climate in the world and in the areas and on any other significant world events that were happening at that time.
  3. Write a fictional account of any part of the exploration that interests you.
  4. Write a personality profile of the major players in the exploration and describe the character of the journey, including the explorers' attitudes toward the cultures of the native peoples.

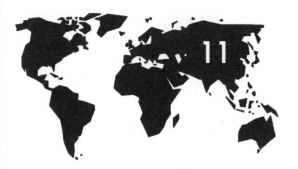

World Exploration

Gertrude Bell

Your Assignment: The Committee has decided that the explorations of Gertrude Bell have a bearing on the success of our mission. Your job is to visit the following sites and report on them. Your report should include one or more of the following:

1. Find the sites listed and explain their importance to the exploration.
2. Draw an annotated map of the journey or journeys; include a key.
3. Report on the results of each journey and how it changed the course of history.
4. Report on any new plant or animal species discovered during the journeys.
5. Report on the cultures that the explorers encountered during their exploration.

36°12′N	37°10′E	33°21′N	44°23′E
54°45′N	1°40′W	33°30′N	36°15′E
46°32′N	8°08′E	27°33′N	41°42′E
33°38′N	42°49′E	31°46′N	35°14′E
32°30′N	43°45′E	37°52′N	32°31′E
34°33′N	38°17′E	30°19′N	35°29′E
35°00′N	38°00′E	35°40′N	51°26′E

Special Instructions: The Committee has been observing your work and has confidence in your abilities and potential. We would like to give you an opportunity for advancement.

- Please investigate the exploration in more depth and include any or all of the following in your report:
    1. Draw and describe in detail the modes of transportation and any other examples of technology that were used on the journeys.
    2. Report on the political climate in the world and in the areas and on any other significant world events that were happening at that time.
    3. Write a fictional account of any part of the exploration that interests you.
    4. Write a personality profile of the major players in the exploration and describe the character of the journey, including the explorers' attitudes toward the cultures of the native peoples.

Gumshoe Geography © 1996 Zephyr Press, Tucson, AZ

World Exploration

Alexandra David-Neel

Your Assignment: The Committee has decided that the explorations of Alexandra David-Neel have a bearing on our mission. Your job is to visit the following sites and report on them. Your report should include one or more of the following:

1. Find the sites listed and explain their importance to the exploration.
2. Draw an annotated map of the journey or journeys; include a key.
3. Report on the results of each journey and how it changed the course of history.
4. Report on any new plant or animal species discovered during the journeys.
5. Report on the cultures that the explorers encountered during their exploration.

22°00′N	98°00′E	35°00′N	105°00′E
14°40′N	17°26′W	44°06′N	6°14′E
38°00′N	137°00′E	29°42′N	91°07′E
27°50′N	88°30′E	44°52′N	12°30′E

Special Instructions: The Committee has been observing your work and has confidence in your abilities and potential. We would like to give you an opportunity for advancement.

- Please investigate the exploration in more depth and include any or all of the following in your report:

1. Draw and describe in detail the modes of transportation and any other examples of technology that were used on the journeys.
2. Report on the political climate in the world and in the areas and on any other significant world events that were happening at that time.
3. Write a fictional account of any part of the exploration that interests you.
4. Write a personality profile of the major players in the exploration and describe the character of the journey, including the explorers' attitudes toward the cultures of the native peoples.

World Exploration

Marco Polo

Your Assignment: The Committee has decided that the explorations of Marco Polo have a bearing on the success of our mission. Your job is to visit the following sites and report on them. Your report should include one or more of the following:

1. Find the sites listed and explain their importance to the exploration.
2. Draw an annotated map of the journey or journeys; include a key.
3. Report on the results of each journey and how it changed the course of history.
4. Report on any new plant or animal species discovered during the journeys.
5. Report on the cultures that the explorers encountered during their exploration.

40°01′N	32°21′E	33°18′N	44°23′E
40°23′N	49°51′E	39°55′N	116°23′W
41°01′N	28°58′E	27°58′N	116°20′E
44°25′N	8°57′E	26°34′N	56°15′E
31°46′N	35°14′E	39°29′N	75°58′E
4°00′N	102°00′E	7°40′N	80°50′E
0°01′N	102°00′E	45°27′N	12°21′E

Special Instructions: The Committee has been observing your work and has confidence in your abilities and potential. We would like to give you an opportunity for advancement.

- Please investigate the exploration in more depth and include any or all of the following in your report:
  1. Draw and describe in detail the modes of transportation and any other examples of technology that were used on the journeys.
  2. Report on the political climate in the world and in the areas and on any other significant world events that were happening at that time.
  3. Write a fictional account of any part of the exploration that interests you.
  4. Write a personality profile of the major players in the exploration and describe the character of the journey, including the explorers' attitudes toward the cultures of the native peoples.

    Gumshoe Geography © 1996 Zephyr Press, Tucson, AZ

World Exploration

Willem Barents, Vitus Bering, Fridtjof Nansen, and Robert Peary

Your Assignment: The Committee has decided that the explorations of Barents, Bering, Nansen, and Peary have a bearing on our mission. Your job is to visit the following sites and report on them. Your report should include one or more of the following:

1. Find the sites listed and explain their importance to the exploration.
2. Draw an annotated map of the journey or journeys; include a key.
3. Report on the results of each journey and how it changed the course of history.
4. Report on any new plant or animal species discovered during the journeys.
5. Report on the cultures that the explorers encountered during their exploration.

74°30'N	19°00'E	65°30'N	169°00'W
78°19'N	72°38'W	81°00'N	55°00'E
90°00'N	0°00'	74°00'N	57°00'E

Special Instructions: The Committee has been observing your work and has confidence in your abilities and potential. We would like to give you an opportunity for advancement.

- Please investigate the exploration in more depth and include any or all of the following in your report:

1. Draw and describe in detail the modes of transportation and any other examples of technology that were used on the journeys.
2. Report on the political climate in the world and in the areas and on any other significant world events that were happening at that time.
3. Write a fictional account of any part of the exploration that interests you.
4. Write a personality profile of the major players in the exploration and describe the character of the journey, including the explorers' attitudes toward the cultures of the native peoples.

World Exploration

Christopher Columbus

Your Assignment: The Committee has decided that the explorations of Christopher Columbus have a bearing on the success of our mission. Your job is to visit the following sites and report on them. Your report should include one or more of the following:

1. Find the sites listed and explain their importance to the exploration.
2. Draw an annotated map of the journey or journeys; include a key.
3. Report on the results of each journey and how it changed the course of history.
4. Report on any new plant or animal species discovered during the journeys.
5. Report on the cultures that the explorers encountered during their exploration.

21°30′N	80°00′W	19°00′N	71°00′W
18°15′N	77°30′W	18°15′N	66°30′W
13°42′N	89°12′W		

Special Instructions: The Committee has been observing your work and has confidence in your abilities and potential. We would like to give you an opportunity for advancement.

- Please investigate the exploration in more depth and include any or all of the following in your report:

1. Draw and describe in detail the modes of transportation and any other examples of technology that were used on the journeys.
2. Report on the political climate in the world and in the areas and on any other significant world events that were happening at that time.
3. Write a fictional account of any part of the exploration that interests you.
4. Write a personality profile of the major players in the exploration and describe the character of the journey, including the explorers' attitudes toward the cultures of the native peoples.

          Gumshoe Geography © 1996 Zephyr Press, Tucson, AZ

World Exploration

Meriwether Lewis and William Clark

Your Assignment: The Committee has decided that the explorations of Lewis and Clark have a bearing on the success of our mission. Your job is to find out about and visit six sites they visited. Your report should include one or more of the following:

1. Find the sites listed and explain their importance to the exploration.
2. Draw an annotated map of the journey or journeys; include a key.
3. Report on the results of each journey and how it changed the course of history.
4. Report on any new plant or animal species discovered during the journeys.
5. Report on the cultures that the explorers encountered during their exploration.

Special Instructions: The Committee has been observing your work and has confidence in your abilities and potential. We would like to give you an opportunity for advancement.

- Please investigate the exploration in more depth and include any or all of the following in your report:

1. Draw and describe in detail the modes of transportation and any other examples of technology that were used on the journeys.
2. Report on the political climate in the world and in the areas and on any other significant world events that were happening at that time.
3. Write a fictional account of any part of the exploration that interests you.
4. Write a personality profile of the major players in the exploration and describe the character of the journey, including the explorers' attitudes toward the cultures of the native peoples.

World Exploration

James Cook

Your Assignment: The Committee has decided that the explorations of James Cook have a bearing on the success of our mission. Your job is to visit the following sites and report on them. Your report should include one or more of the following:

1. Find the sites listed and explain their importance to the exploration.
2. Draw an annotated map of the journey or journeys; include a key.
3. Report on the results of each journey and how it changed the course of history.
4. Report on any new plant or animal species discovered during the journeys.
5. Report on the cultures that the explorers encountered during their exploration.

10°30'S 105°40'E	41°20'S 174°25'E
27°07'S 109°22'W	19°10'S 149°00'E
24°00'N 167°00'W	17°37'S 149°27'W
20°00'S 175°00'W	

Special Instructions: The Committee has been observing your work and has confidence in your abilities and potential. We would like to give you an opportunity for advancement.

- Please investigate the exploration in more depth and include any or all of the following in your report:

  1. Draw and describe in detail the modes of transportation and any other examples of technology that were used on the journeys.
  2. Report on the political climate in the world and in the areas and on any other significant world events that were happening at that time.
  3. Write a fictional account of any part of the exploration that interests you.
  4. Write a personality profile of the major players in the exploration and describe the character of the journey, including the explorers' attitudes toward the cultures of the native peoples.

Gumshoe Geography © 1996 Zephyr Press, Tucson, AZ

World Exploration

Robert Burke, William Wills, and John Stuart

Your Assignment: The Committee has decided that the explorations of Burke, Wills, and Stuart have a bearing on our mission. Your job is to visit the following sites and report on them. Your report should include one or more of the following:

1. Find the sites listed and explain their importance to the exploration.
2. Draw an annotated map of the journey or journeys; include a key.
3. Report on the results of each journey and how it changed the course of history.
4. Report on any new plant or animal species discovered during the journeys.
5. Report on the cultures that the explorers encountered during their exploration.

32°42′S	26°20′E	28°29′S	137°46′E
12°28′S	130°50′E	17°44′S	139°22′E
37°49′S	144°58′E	32°24′S	142°26′E
25°00′S	137°00′E		

Special Instructions: The Committee has been observing your work and has confidence in your abilities and potential. We would like to give you an opportunity for advancement.

- Please investigate the exploration in more depth and include any or all of the following in your report:

  1. Draw and describe in detail the modes of transportation and any other examples of technology that were used on the journeys.
  2. Report on the political climate in the world and in the areas and on any other significant world events that were happening at that time.
  3. Write a fictional account of any part of the exploration that interests you.
  4. Write a personality profile of the major players in the exploration and describe the character of the journey, including the explorers' attitudes toward the cultures of the native peoples.

World Exploration

Ferdinand Magellan and Sebastian del Cano

Your Assignment: The Committee has decided that the explorations of Magellan and del Cano have a bearing on our mission. Your job is to visit the following sites and report on them. Your report should include one or more of the following:

1. Find the sites listed and explain their importance to the exploration.
2. Draw an annotated map of the journey or journeys; include a key.
3. Report on the results of each journey and how it changed the course of history.
4. Report on any new plant or animal species discovered during the journey.
5. Report on the cultures that the explorers encountered during their exploration.

28°00'N	15°30'W	34°21'S	18°28'E
9°00'N	168°00'E	2°00'S	128°00'E
44°00'S	68°00'W	13°00'N	122°00'E
40°00'N	4°00'W	54°00'S	69°00'W

Special Instructions: The Committee has been observing your work and has confidence in your abilities and potential. We would like to give you an opportunity for advancement.

- Please investigate the exploration in more depth and include any or all of the following in your report:

1. Draw and describe in detail the modes of transportation and any other examples of technology that were used on the journeys.
2. Report on the political climate in the world and in the areas and on any other significant world events that were happening at that time.
3. Write a fictional account of any part of the exploration that interests you.
4. Write a personality profile of the major players in the exploration and describe the character of the journey, including the explorers' attitudes toward the cultures of the native peoples.

Gumshoe Geography © 1996 Zephyr Press, Tucson, AZ

World Exploration

First People on the Mountain

Your Assignment: The Committee has decided that the early conquest of the following mountains has a bearing on our mission. Your job is to visit the following sites and report on them. Your report should include one or more of the following:

1. Find the sites listed and explain their importance to the exploration.
2. Draw an annotated map of the journey or journeys; include a key.
3. Report on the results of each journey and how it changed the course of history.
4. Report on any new plant or animal species discovered during the journey.
5. Report on the cultures that the explorers encountered during their exploration.

32°39′S	70°00′W	45°50′N	6°52′E
10°50′N	73°45′W	27°59′N	86°56′E
35°53′N	76°30′E	3°04′S	37°22′E
38°51′N	105°03′W	27°08′S	109°26′W
63°30′N	151°00′W	19°02′N	98°38′W
78°35′S	85°25′W	45°58′N	7°39′E

Special Instructions: The Committee has been observing your work and has confidence in your abilities and potential. We would like to give you an opportunity for advancement.

■ Please investigate the exploration in more depth and include any or all of the following in your report:

1. Draw and describe in detail the modes of transportation and any other examples of technology that were used on the journeys.
2. Report on the political climate in the world and in the areas and on any other significant world events that were happening at that time.
3. Write a fictional account of any part of the exploration that interests you.
4. Write a personality profile of the major players in the exploration and describe the character of the journey, including the explorers' attitudes toward the cultures of the native peoples.

1
ONE COUNTRY

Teacher's Guide

Teacher's Guide
One Country 1-1

Site: India
Assignment Focus: Site feasibility

Students should make recommendations for placement of transmission towers in rural or urban setting, looking for details of cultural and environmental data and possible risk factors.

23°02'N	72°37'E	Ahmadabad
18°58'N	72°50'E	Bombay
22°32'N	88°22'E	Calcutta
17°23'N	78°28'E	Hyderabad
16°56'N	82°13'E	Kakinada
31°35'N	74°18'E	Lahore
28°36'N	77°12'E	New Delhi
25°36'N	85°07'E	Patna
34°05'N	74°49'E	Srinagar
8°29'N	76°55'E	Trivandrum

Special Instructions Objective: Relations with neighbors

Students are looking for information about the governments of India's neighbors along its borders and indications from current events and recent history about how the countries feel about one another.

Teacher's Guide
One Country 1-2

Site: Western Canada
Assignment Focus: Cities in western Canada

Students should research mineral deposits in western Canada and find out how the deposits are mined, transported, refined, and marketed.

62°05'N	136°18'W	Carmacks, Yukon
58°46'N	94°10'W	Churchill, Manitoba
64°04'N	139°25'W	Dawson, Yukon
54°56'N	101°53'W	Flin Flon, Manitoba
50°40'N	120°20'W	Kamloops, BC
54°05'N	128°38'W	Kitimat, BC
50°03'N	110°40'W	Medicine Hat, Alberta
50°23'N	105°23'W	Moose Jaw, Saskatchewan
54°01'N	124°01'W	Vanderhoof, BC
62°27'N	114°21'W	Yellowknife, NWT

Special Instructions Objective: Recreation

Students should decide on the characteristics of a good recreational facility and then match several of the areas in western Canada to the list of characteristics. Individual areas may be suited to particular recreational activities. These activities may be categorized by difficulty, energy, and skill requirements.

Teacher's Guide
One Country 1-3

Site: Eastern Canada
Assignment Focus: French/English/Indian relations

Students should research the current status and the history of the present situation. Students might be encouraged to make suggestions for ways to solve any conflicts and to improve relations.

48°00'N	66°40'W	Campbellton, NB
49°53'N	74°21'W	Chibougamau, Québec
48°26'N	71°04'W	Chicoutimi, Québec
53°50'N	79°00'W	Chisasibi, Québec
49°47'N	92°50'W	Dryden, Ontario
48°57'N	54°34'W	Gander, Newfoundland
51°17'N	80°39'E	Moosonee, Ontario
57°00'N	61°40'W	Nain, Labrador Highlands
46°09'N	60°11'W	Sydney, Nova Scotia
48°23'N	89°15'W	Thunder Bay, Ontario

Special Instructions Objective: Electrical power

Students should find information on the damming of the rivers that flow into Hudson Bay. They should be able to find how it has affected the way of life of the American Indian population in that area.

Teacher's Guide
One Country 1-4

Site: Canadian Fort Cities
Assignment Focus: Cities that are named after forts

Students should be able to find information about the early history of Canada and about the establishment of frontier forts.

52°15'N	81°37'W	Fort Albany, Ontario
71°37'N	117°57'W	Fort Collinson, NWT
65°12'N	123°26'W	Fort Franklin, NWT
49°43'N	113°25'W	Fort Macleod, Alberta
58°49'N	122°39'W	Fort Nelson, BC
62°42'N	109°08'W	Reliance, NWT
53°43'N	113°13'W	Fort Saskatchewan, AL
61°52'N	121°23'W	Fort Simpson, NWT
56°15'N	120°15'W	Fort St. John, BC
58°24'N	116°00'W	Fort Vermillion, Alberta

Special Instructions Objective: Migration patterns

Students should be able to find several mammal and bird migratory routes in this area. Students should produce attractive maps showing the migration patterns as well as an information sheet about each species.

Teacher's Guide
One Country 1-5

Site: Coastal Mexico
Assignment Focus: Mexican resort towns

Students should research the resort towns along the coasts of Mexico and compare the quality and variability of the offerings of each resort. Students should be encouraged to state preferences and be able to justify their choices.

16°51'N	99°55'W	Acapulco
19°14'N	103°43'W	Colima
19°51'N	90°32'W	Campeche
21°05'N	86°46'W	Cancún
27°59'N	109°56'W	Cuidad Obregón
24°10'N	110°18'W	La Paz
23°13'N	106°25'W	Mazatlán
20°58'N	89°37'W	Mérida
22°13'N	97°51'W	Tampico
19°12'N	96°08'W	Veracruz

Special Instructions Objective: Marine vocations

Students should find information on the many different ways that people make a living from the sea in Mexico. They should look at equipment and practices and even how the people of the area have brought the sea into their folklore and culture.

Teacher's Guide
One Country 1-6

Site: Interior Mexico
Assignment Focus: Desert plants and animals

Students should study and report on the ecosystem and food webs in the mountain-desert region of central Mexico. Students should also find out how people live in this region and make suggestions for low-impact improvements in the way of life.

21°53'N	102°18'W	Aguascalientes
28°30'N	106°00'W	Chihuahua
27°37'N	100°44'W	Ciudad Juárez
20°40'N	103°20'W	Guadalajara
20°58'N	89°37'W	Mérida
25°41'N	100°19'W	Monterrey
17°00'N	96°30'W	Oaxaca
19°19'N	98°14'W	Tlaxcala
25°33'N	103°26'W	Torreón
16°45'N	93°07'W	Tuxtla Gutiérrez
24°02'N	104°40'W	Victoria de Durango

Special Instructions Objective: Desert travel

Students should study the natural environment and climate of Central Mexico, make a list of desirable characteristics for a vehicle, and then design one that would meet the criteria and still be practical.

Teacher's Guide
One Country 1-7

Site: Argentina
Assignment Focus: Food exports

Students should be able to find information about food goods that are produced in Argentina and shipped all over the world. They should be able to classify the goods and determine if there are any goods that are considered to be world class.

38°44'S	62°16'W	Bahía Blanca
34°36'S	58°27'W	Buenos Aires
45°50'S	67°30'W	Comodoro Ravadavia
31°25'S	64°10'W	Córdoba
34°55'S	57°57'W	La Plata
32°54'S	68°50'W	Mendoza
51°37'S	69°10'W	Río Gallegos
33°20'S	66°20'W	San Luis
26°49'S	65°13'W	San Miguel de Tucumán
27°50'S	64°15'W	Santiago del Estero

Special Instructions Objective: Local vocations

The focus of this activity is to identify the regionally unique vocations of the rural populace and to compare those with regionally unique vocations in the United States.

Teacher's Guide
One Country 1-8

Site: Colombia
Assignment Focus: Precious minerals

Students should find information on the jewel-mining industry and trade in Colombia. Emeralds are most plentiful here, but other jewels are also mined in this region.

4°36'N	74°05'W	Bogotá
7°08'N	73°09'W	Bucaramanga
3°53'S	77°04'W	Buenaventura
10°25'N	75°32'W	Cartagena
7°54'N	72°31'W	Cúcuta
6°15'N	75°35'W	Medellín
1°08'N	70°03'W	Mitú
1°13'N	77°17'W	Pasto
2°27'N	76°36'W	Popayán
11°15'N	74°13'W	Santa Marta

Special Instructions Objective: Life in a seismic and volcanic region

Students should find out how the rural population adapts to the frequent earthquakes and volcanic activity in their region. They should speculate on ways that the native population could improve their coping strategies.

Teacher's Guide
One Country 1-9

Site: Poland
Assignment Focus: Industry and manufacturing
Students should find information about the various industries in Poland, the resources available, and the advantages of location, climate, and environment.

53°08'N	18°00'E	Bydgoszcz
51°10'N	23°28'E	Chelm
54°23'N	18°40'E	Gdansk
54°12'N	15°33'E	Kolobrzeg
50°03'N	19°58'E	Krakow
51°46'N	19°30'E	L¢dz
51°15'N	22°35'E	Lublin
53°24'N	14°32'E	Szczecin
52°15'N	21°00'E	Warsaw
51°06'N	17°00'E	Wroclaw

Special Instructions Objective: History of conquest
Poland has been the site of many conquests and different governments, partly because of its strategic location and the lack of natural borders. Students should be able to find examples of change and explain why they took place.

Teacher's Guide
One Country 1-10

Site: United Kingdom
Assignment Focus: Historical buildings and monuments
Students should find information on monuments and historical sites in or near the designated cities in the United Kingdom. They should be able to date them and report on their historical significance.

54°35'N	5°55'W	Belfast
52°30'N	1°50'W	Birmingham
51°30'N	3°13'W	Cardiff
51°54'N	8°28'W	Cork
53°20'N	6°15'W	Dublin
56°28'N	3°00'W	Dundee
55°53'N	4°15'W	Glasgow
57°27'N	4°15'W	Inverness
53°25'N	2°55'W	Liverpool
51°30'N	0°10'W	London
52°58'N	1°10'W	Nottingham
50°23'N	4°10'W	Plymouth

Special Instructions Objective: Cultural heritage
Students should be able to find information about Welsh, Scottish, Irish, and English cultural heritages and report on unique customs that people still observe. They should be able to see that it is possible and even desirable to have cultural diversity within a national structure.

Teacher's Guide
One Country 1-11

Site: Japan
Assignment Focus: Energy needs and consumption
Students should find information on the sources and types of energy used in Japan. They should be able to tell where the energy fuels come from, how expensive they are, and how they influence Japan's role in world politics.

40°49'N	140°45'E	Aomori
43°46'N	140°22'E	Asahikawa
42°45'N	140°43'E	Hakodate
34°24'N	132°27'E	Hiroshima
31°36'N	130°33'E	Kagoshima
32°47'N	129°56'E	Nagasaki
37°55'N	139°03'E	Niigata
35°37'N	137°14'E	Osaka
43°03'N	141°21'E	Sapporo (Otaru)
38°15'N	140°53'E	Sendai
34°21'N	134°03'E	Takamatsu
35°40'N	139°46'E	Tokyo

Special Instructions Objective: Family structure
Students should be able to report on the role of the traditional family in Japanese life and about the changes that are taking place because of Western influence. They should form an opinion about the negative and positive aspects of the changes.

Teacher's Guide
One Country 1-12

Site: Eastern Russia
Assignment Focus: Shortages of necessities
Students should find information about how the Russian people are suffering and coping as a result of the breakup of the Soviet Union and the change in the government's philosophy and focus.

69°25'N	86°15'E	Dudinka
48°27'N	135°06'E	Habarovsk
62°13'N	129°49'E	Jakutsk
56°01'N	92°50'E	Krasnojarsk
59°34'N	150°48'E	Magadan
55°02'N	82°55'E	Novosibirsk
53°34'N	142°56'E	Oha
55°00'N	73°24'E	Omsk
54°54'N	69°06'E	Petropavlovsk
56°30'N	84°58'E	Tomsk
66°13'N	169°48'W	Uelen
43°10'N	131°56'E	Vladivostok

Special Instructions Objective: National strength
Students should find information about invasions of Russia and about how the Russian people have been able to frustrate and repel invaders. They should form an opinion about the factors that contribute to this phenomenon.

Teacher's Guide
One Country 1-13

Site: Western Russia
Assignment Focus: Siberian natural environment

Students should find information about the natural environment of Siberia. The information should include the climate and nature of the forest and the species of animals that live there. They should apprise themselves of how accessible the information should be.

64°34'N	40°32'E	Archangelsk
46°21'N	48°03'E	Astrahan
55°55'N	37°57'E	Kaliningrad
55°45'N	49°08'E	Kazan
61°16'N	46°35'E	Kotlas
55°45'N	37°35'E	Moscow
68°58'N	33°05'E	Murmansk
58°00'N	56°15'E	Perm
51°34'N	46°02'E	Saratov
59°55'N	30°15'E	St. Petersburg
61°40'N	50°46'E	Syktyvkar
48°44'N	44°25'E	Volgograd

Special Instructions Objective: Community isolation

Most of Siberia is a vast tangled forest, and the communities there are fairly isolated. Students should find out how people get their essential goods and how they stay in touch with the rest of the world. They should also determine whether the quality of life should be improved for these people and if so how it could be.

Teacher's Guide
One Country 1-14

Site: North and South Korea
Assignment Focus: Comparative governments

Students should find information about the governments of North and South Korea, their history, and how each government views itself in terms of world affairs.

41°46'N	129°49'E	Ch'oujin
39°50'N	127°38'E	Hungnam
42°24'N	128°10'E	Hyesan
37°28'N	126°38'E	Inch'on
40°58'N	126°36'E	Kanggye
35°09'N	126°55'E	Kwangju
38°25'N	127°17'E	P'yînggang
35°06'N	129°03'E	Pusan
37°34'N	127°00'E	Seoul
39°10'N	127°26'E	Wonsan

Special Instructions Objective: Comparative standard of living

Students should be able to compare the ways of life and the standards of living of the two countries and make an assessment of the quality of life in each of the two cultures.

Teacher's Guide
One Country 1-15

Site: Turkey
Assignment Focus: Native dwellings and shelter

Students should find information about the way housing is constructed in Turkey. They should find information on traditional as well as contemporary shelter and be able to explain the reasons for each kind.

37°01'N	35°18'E	Adana
40°46'N	30°24'E	Adapazari
39°56'N	32°52'E	Ankara
37°55'N	40°14'E	Diyarbakir
37°05'N	37°22'E	Gaziantep
41°01'N	28°58'E	Istanbul
38°25'N	27°09'E	Izmir
38°21'N	38°19'E	Malatya
40°59'N	39°43'E	Trabzon
38°28'N	43°20'E	Van

Special Instructions Objective: European-Asian mix

Turkey is a rugged land between Asia and Europe. Students should be able to document European as well as Asian influence in the culture.

Teacher's Guide
One Country 1-16

Site: Indonesia
Assignment Focus: Rain forest products

Students should find information about the oil, rubber, and palm products industries. They should also determine from research how the rain forest environment is being treated and if there is a long-term growth or short-term profit strategy for these industries.

3°43'S	128°12'E	Ambon, Ceram
1°17'S	116°50'E	Balikpapau, Borneo
6°10'S	106°46'E	Jakarta, Java
1°29'N	124°51'E	Manado, Celebes
3°35'N	98°40'E	Medan, Sumatra
0°57'S	100°21'E	Padang, Sumatra
0°02'S	109°20'E	Pontianak, Borneo
7°15'S	112°45'E	Surabaya, Java

Special Instructions Objective: Overcrowded vs uninhabited sites

Indonesia includes Java, with the highest population density in the world, and its next-door neighbor Sumatra, which is nearly uninhabited. Students should research the factors that might lead to this situation, determine its causes, and make suggestions for changes.

Teacher's Guide
One Country 1-17

Site: Zaire
Assignment Focus: Minerals and raw materials
Students should find information about the copper and cobalt mines in the southern part of Zaire. It is said that Solomon's mines are in the southern region of Central Africa. Students might be encouraged to find out about this possibility.

3°18'S	17°20'E	Bandundu
2°30'S	28°52'E	Bukavu
4°18'S	15°18'E	Kinshasa
0°25'N	25°12'E	Kisangani
2°09'N	21°31'E	Lisala
11°40'N	27°30'E	Lubumbashi
5°49'S	13°27'E	Matadi
0°04'N	18°16'E	Mbandaka

Special Instructions Objective: Jungle legends
Deepest, darkest Africa is found in Zaire or what was once known as the Belgian Congo. Students should find legends about the African jungle and report on them. They should be able to sort truth from fable. An additional activity is to have students write original fables based on fact.

Teacher's Guide
One Country 1-18

Site: South Africa
Assignment Focus: Precious mineral wealth
Students should find information about how gold and diamonds were discovered, how the industry grew, who ran it, who controlled it, and who worked it. They should know its history and its present state.

29°12'S	26°07'E	Bloemfontein
35°55'S	18°22'E	Cape Town
29°55'S	30°56'E	Durban
33°00'S	27°55'E	East London
28°43'S	24°46'E	Kimberley
23°54'S	29°25'E	Pietersburg
33°58'S	25°40'E	Port Elizabeth
25°45'S	28°10'E	Pretoria
28°47'S	32°06'E	Richards Bay
28°25'S	21°15'E	Upington

Special Instructions Objective: Racial conflict
Since the seventh century, South Africa has been considered a strategic site, first by the Dutch, then by the English. The discovery of diamonds and gold really heated things up. Students should find out about the region's history and how the local population has been affected. The information should be compiled into a cause-and-effect time line.

Teacher's Guide
One Country 1-19

Site: Madagascar
Assignment Focus: Rain forest misuse
Students should find information about the forest industry in Madagascar and how the relatively short-sighted forest practices have changed the environment of the island. Students should be able to tell how the population as well as the ecosystem have been affected by this change.

24°95'S	44°04'E	Androka
19°00'S	46°40'E	Antananarivo
12°17'S	49°17'E	Antseranana
19°51'S	47°01'E	Antsirabe
21°28'S	47°05'E	Fianarantsoa
15°17'S	46°43'E	Mahajanga
18°10'S	49°24'E	Toamasina
23°21'S	43°39'E	Toliary

Special Instructions Objective: Unique, isolated species
The isolation of the island of Madagascar has led to the microevolution of many unique plant and animal species. Students should be able to report on the various species and how they are adapted to their distinct niches.

Teacher's Guide
One Country 1-20

Site: Italy
Assignment Focus: Art, music, and exploration
Students should find information about developments in the arts and in exploration and discovery during the period following the Middle Ages. This is a good chance to delve into biographical data.

43°38'N	13°30'E	Ancona
41°08'N	16°51'E	Bari
46°31'N	11°22'E	Bolzano
37°30'N	15°06'E	Catania
44°25'N	8°57'E	Genoa
45°28'N	9°12'E	Milan
40°50'N	14°15'E	Naples
38°07'N	13°22'E	Palermo
42°45'N	12°29'E	Rome
40°28'N	17°14'E	Taranto
45°40'N	13°46'E	Trieste
45°27'N	12°21'E	Venice

Special Instructions Objective: Microregional diversity
Italy has many small and distinct regions with their own subcultures. Students should create a chart identifying and categorizing several of the regions by the factors that are listed on the student work sheet.

Teacher's Guide
One Country 1-21

Site: France
Assignment Focus: Agricultural practices

Next to Russia, France is Europe's largest producer of food. Students should research agricultural products and practices in France. They should note how the environment and climate affect the products and methods.

44°50'N	0°34'W	Bordeaux
48°42'N	4°29'W	Brest
48°00'N	0°12'E	Le Mans
45°45'N	4°51'E	Lyon
43°18'N	5°24'E	Marseilles
47°13'N	1°33'W	Nantes
43°42'N	7°15'E	Nice
48°52'N	2°20'E	Paris
42°41'N	2°53'E	Perpignan
49°15'N	4°02'E	Reims
49°26'N	1°05'E	Rouen
48°35'N	7°45'E	Strasbourg

Special Instructions Objective: History of leadership

France's past is brightly colored by its leaders from Roman times to the present. Students should study the leaders and decide which ten influenced the destiny of France the most, positively or negatively. They should then enter short biographies on a time line of France's history.

Teacher's Guide
One Country 1-22

Site: New Zealand
Assignment Focus: Member's choice

Students should choose an area of interest in reporting about New Zealand. Possible areas include the sheep population, the Alps, the Maoris, geothermal energy, unique species, the effects of the proximity to the Antarctic and the ozone hole, and so on.

36°52'S	174°45'E	Auckland
43°32'S	172°37'E	Christchurch
45°53'S	170°30'E	Dunedin
38°39'S	178°01'E	Gisborne
37°47'S	175°17'E	Hamilton
46°25'S	168°21'E	Invercargill
41°16'S	173°15'E	Nelson
38°09'S	176°15'E	Rotorua
44°24'S	171°15'E	Timaru
41°17'S	174°46'E	Wellington

Special Instructions Objective: Educational system

New Zealand is the home of many learning and teaching innovations. Students could learn about the educational system from primary through college. They should be able to recognize any exemplary programs or methods and to be able to explain worth and success of those programs and methods.

Teacher's Guide
One Country 1-23

Site: Chile
Assignment Focus: Domesticated animals

Students should find information about the animals that people use for food and work. They should find out about the origins of the animals, how they were domesticated, and what they are used for.

23°39'S	70°24'W	Antofagasta
18°29'S	70°20'W	Arica
36°50'S	73°03'W	Concepción
41°28'S	72°57'W	Puerto Montt
51°44'S	72°31'W	Puerto Natales
53°09'S	70°55'W	Punta Arenas
34°35'S	71°00'W	San Fernando
33°27'S	70°40'W	Santiago
39°48'S	73°14'W	Valdivía
33°02'S	71°38'W	Valparaiso

Special Instructions Objective: Andean music

Chile has a strong traditional music heritage. The traditional instruments are quite distinctive. Students should be able to find information about the music as well as examples of it. They might even be able to obtain Chilean instruments and learn to play simple melodies on them.

Teacher's Guide
One Country 1-24

Site: Malaysia
Assignment Focus: Tin mining and export

Students should find information about the history and development of the tin industry in Malaysia. Comparisons should be made to other tin-producing regions of the world.

3°10'N	113°02'E	Bintulu
5°25'N	100°20'E	George Town
5°59'N	116°04'E	Kota Kinabula
3°10'N	101°42'E	Kuala Lumpur
1°33'N	110°20'E	Kuching Sarawak
5°50'N	118°07'E	Sandakan
2°18'N	111°49'E	Sibu
1°17'N	103°51'E	Singapore

Special Instructions Objective: Rice production

Students should look at Malaysia as a food-producing country and find out about rice production there. They should be able to compare it with other rice-producing regions of the world. They should also be able to comment on how the rice crop regulates the lives of those who cultivate it.

Teacher's Guide
One Country 1-25

Site: China
Assignment Focus: The early silk and spice trade
Students should find information about the history of trade with China, from the time of Marco Polo to the present. They should be able to report on regional goods and resources that are available today in that region of the world.

39°55'N	116°23'E	Beijing
43°51'N	125°20'E	Changchun
30°45'N	104°04'E	Chengdu
45°45'N	126°37'E	Harbin
25°08'N	102°43'E	Kunming
29°42'N	91°07'E	Lhasa
31°59'N	118°51'E	Nanjing
31°41'N	121°28'E	Shanghai
43°48'N	87°35'E	Örumqi
34°15'N	108°50'E	Xi'an

Special Instructions Objective: Transportation
Students should find information about modes of transportation in China. They should be able to report on the development of modern means of travel and compare them with those in other nations. They should be able to make recommendations for change and improvement.

Teacher's Guide
One Country 1-26

Site: Australia
Assignment Focus: Water shortages
Students should find information about the history and development of water resources in Australia. They should report on all facets of water use and production.

34°56'S	138°36'E	Adelaide
23°42'S	133°53'E	Alice Springs
27°28'S	153°02'E	Brisbane
17°58'S	122°14'E	Broome
12°28'S	130°50'E	Darwin
28°46'S	114°36'E	Geraldton
30°45'S	121°28'E	Kalgoorlie-Boulder
37°49'S	144°58'E	Melbourne
20°44'S	139°30'E	Mount Isa
31°56'S	115°50'E	Perth
33°52'S	151°13'E	Sydney
19°16'S	146°48'E	Townsville

Special Instructions Objective: Cultural roots
Students should find information about the use of Australia as a penal colony and about how those prisoners settled and colonized the country. They should also research the way of life of the Aborigines and how the two cultures relate to each other.

Teacher's Guide
One Country 1-27

Site: Coastal Brazil
Assignment Focus: Resorts, Iberian heritage
Students should find out about the recreational opportunities in or near the coastal cities of Brazil and about the Spanish and Portuguese influences that can be found there. They should report on how people make a living and how their Iberian heritage affects their way of life.

1°27'S	48°29'W	Belém
3°43'S	38°30'W	Fortaleza
5°47'S	35°13'W	Natal
30°04'S	51°11'W	Pôrto Alegre
8°03'S	34°54'W	Recife
22°54'S	43°15'W	Rio de Janeiro
12°59'S	38°31'W	Salvador
23°57'S	46°20'W	Santos
2°31'S	44°16'W	São Luís

Special Instructions Objective: Communication
Students should find out how the people of coastal Brazil communicate with one another and with the rest of the world. They should report on any innovative or outdated ways that people use to keep in touch and should make recommendations for improvement or adaptation.

Teacher's Guide
One Country 1-28

Site: Interior Brazil
Assignment Focus: Life expectancy
Students should report on statistics about Brazil's birth rate, death rate, infant mortality rate, life expectancy, and so on. They should speculate about the factors that affect these statistics and about ways to improve the situation.

19°55'S	43°56'W	Belo Horizonte
15°47'S	47°55'W	Brasília
20°27'S	60°50'W	Campo Grande
19°01'S	57°39'S	Corumbá
15°35'S	56°05'W	Cuiabá
25°25'S	49°15'W	Curitiba
5°50'N	55°10'W	Paramaribo
8°46'S	63°45'W	Pôrto Velho
9°58'S	67°48'W	Rio Branco
29°41'S	53°48'W	Santa Maria
23°32'S	46°37'W	São Paulo
19°45'S	47°55'W	Uberaba

Special Instructions Objective: Dependance on rivers
Students should look at how the interior areas of Brazil are served by the network of rivers that feed the Amazon and how the populace uses the rivers. They should report on any other resources that influence the economy or daily life.

Teacher's Guide
One Country 1-29

Site: Germany
Assignment Focus: North vs south
Students should find information contrasting the
industrial north of Germany with the rural, agricultural,
and forested south. They should be able to report
about the way of life in each region and also about
what things help the standard of living in each region.

52°31'N	13°24'E	Berlin
50°44'N	7°06'E	Bonn
53°05'N	8°48'E	Bremen
51°03'N	13°45'E	Dresden
50°55'N	13°22'E	Freiburg
52°22'N	9°43'E	Hannover
51°18'N	12°20'E	Leipzig
48°09'N	11°35'E	Munich
49°27'N	11°05'E	Hamburg
54°05'N	12°08'E	Rostock
48°46'N	9°11'E	Stuttgart
50°05'N	8°15'E	Wiesbaden

Special Instructions Objective: German culture
Students should look at all aspects of life in the different cultural regions of Germany and report on the
commonalities. They should be able to create a chart
showing similarities and differences among regions.

2
WORLD REGIONS

Teacher's Guide

Teacher's Guide
World Regions 2-1

Site: Scandinavia
Assignment Focus: Glacial topography
Students should find out about the natural environment of Scandinavia, particularly in terms of glaciers and the landforms that they created.

57°03'N	9°56'E	Elborg, Denmark
60°23'N	5°20'E	Bergen, Norway
55°40'N	12°35'E	Copenhagen, Denmark
57°43'N	11°58'E	Gîteborg, Sweden
60°01'N	24°58'E	Helsinki, Finland
68°58'N	33°05'E	Murmansk, Russia
64°30'N	11°30'E	Namsos, Norway
59°55'N	10°45'E	Oslo, Norway
64°09'N	21°57'W	Reykjavik, Iceland
59°20'N	18°03'E	Stockholm, Sweden
69°40'N	19°00'E	Tromso, Norway
63°06'N	21°36'E	Vaasa, Finland

Special Instructions Objective: Standard of living
Students should investigate and report on the factors that facilitate the high standard of living in Scandinavia. They should be able to speculate about cause-and-effect relationships, and they should be able to make suggestions about how to transfer certain factors to other cultures.

Teacher's Guide
World Regions 2-2

Site: Balkan Peninsula
Assignment Focus: Early governments
Students should research the philosophies and implementation of ancient Roman and Greek forms of government and how they affected the way of life in this region at that time. Students could also report on other forms of government in that area.

37°59'N	23°44'E	Athens, Greece
37°02'N	22°07'E	Kalámai, Greece
39°40'N	19°45'E	Kérkira, Greece
40°37'N	20°46'E	Koráa, Albania
41°21'N	21°34'E	Prilep, Macedonia
42°05'N	19°30'E	Shkodra, Albania
42°00'N	21°29'E	Skopje, Macedonia
40°38'N	22°56E	Thessaloniki, Greece
41°20'N	19°50'E	Tirana, Albania
41°08'N	24°53'E	Xanthi, Greece

Special Instructions Objective: Languages and their roots
Students should find out about modern languages spoken in the Balkan region and relate those languages to their roots. Students should report on how the languages have been standardized and also about any regional dialects that are prominent.

Teacher's Guide
World Regions 2-3

Site: Bavaria
Assignment Focus: Alpine transportation
Students should find out about the kinds of transportation that tie the region together. They should also find out how these forms of transportation have developed and how the local populace views the private versus public transportation.

46°31'N	11°22'E	Bolzano, Italy
46°10'N	6°10'E	Geneva, Switzerland
47°16'N	11°24'E	Innsbruck, Austria
48°18'N	14°18'E	Linz, Austria
46°05'N	8°20'E	Luzern, Switzerland
48°09'N	11°35'E	Munich, Germany
47°48'N	13°02'E	Salzburg, Austria
47°08'N	9°30'E	Vaduz, Leichtenstein
48°12'N	16°22'E	Vienna, Austria
47°20'N	8°35'E	Zurich, Switzerland

Special Instructions Objective: Community isolation
Students should investigate the dairy industry in the region, especially as it relates to the production of local cheeses. Other food products and crops that are distinctive to the region should be reported on also.

Teacher's Guide
World Regions 2-4

Site: Iberia
Assignment Focus: Cork production
Students should do research about the cork industry, from the harvest to export and world uses and consumption.

41°23'N	2°11'E	Barcelona, Spain
36°32'N	6°18'W	Cádiz, Spain
37°36'N	0°59'W	Cartegena, Spain
38°34'N	7°54'W	Êvora, Portugal
38°43'N	9°08'W	Lisbon, Portugal
40°24'N	3°41'W	Madrid, Spain
36°43'N	4°25'W	Málaga, Spain
41°09'N	8°37'W	Porto, Portugal
43°19'N	1°59'W	San Sebastian, Spain
37°23'N	5°59'W	Seville, Spain
39°50'N	4°00'W	Toledo, Spain
39°28'N	0°22'W	Valencia, Spain
41°35'N	1°00'W	Zaragoza, Spain

Special Instructions Objective: Local health issues
Students should report on health issues in this agricultural region. They should speculate as to how the population's health is affected by the lack of heavy industry in the region. They should find other factors that directly affect people's health.

Teacher's Guide
World Regions 2-5

Site: North Africa
Assignment Focus: Growth of desert environment
Students should report on the research about the solutions and methods to slow or stop the encroachment of the desert onto arable land in the North Africa/Sahara region. They should also determine how much of a problem it is to the local people and how these people view the solutions.

24°12'N	23°18'E	Al Jauf, Libya
36°47'N	3°03'E	Algiers, Algeria
24°05'N	32°53'E	Aswan, Egypt
17°55'N	19°07'E	Faya-Largeau, Chad
19°04'N	8°24'E	Iferouâne, Niger
31°38'N	8°00'W	Marrakech, Morocco
20°45'N	17°01'W	Nouadhibou, Mauritania
33°53'N	10°07'E	Qabis, Tunisia
22°42'N	3°56'W	Taoudenni, Mali
27°45'N	8°25'W	Tindouf, Algeria

Special Instructions Objective: Trade/Currency
Students should find out about trade and currency in this region. They should be able to research the history of trade and currency and chart their development over time. Students should be able to comment on whether world trade has had a standardizing effect on the region's trade and currency.

Teacher's Guide
World Regions 2-6

Site: Northern South America
Assignment Focus: Life in a mountainous region
Students should find out what life is like in the northern Andes region of South America. They should be able to map or explain how extensive the mountain cover is and tell how people meet basic needs in this region.

10°30'N	66°56'W	Caracas, Venezuela
4°56'N	52°20'W	Cayenne, French Guiana
8°08'N	63°33'W	Ciudad Bolivar, Venezuela
6°48'N	58°10'W	Georgetown, Guyana
10°40'N	71°37'W	Maracaibo, Venezuela
9°45'N	63°11'W	Maturín, Venezuela
5°50'N	55°10'W	Paramaribo, Surinam
1°55'N	67°04'W	San Carlos de Río Negro
7°46'N	72°14'W	San Cristóbal, Venezuela
10°11'N	67°45'W	Valencia, Venezuela

Special Instructions Objective: Inca influence
Students should look at the way of life in ancient Incan society and report on customs and daily life. They should be able to find out about the factors that caused the Incas' decline. Students should also be able to comment on the influence of the Inca culture on modern culture in this region.

Teacher's Guide
World Regions 2-7

Site: Oceania
Assignment Focus: International dateline
Students should report on the foundation of the international dateline and be able to report on how it affects travel in the Pacific region. They should also speculate how it affects people who live near it.

25°04'S	130°05'W	Adamstown, Pitcairn Island
13°28'N	144°45'E	Agana, Guam
21°11'S	159°46'W	Avarua, Cook Islands
1°20'N	173°01'E	Bairiki, Kiribati
9°27'S	159°57'E	Honiara, Soloman Islands
21°19'N	157°52'W	Honolulu, Hawaii
7°20'N	134°30'E	Koror, Palau
22°16'S	166°26'E	Nouméa, New Caledonia
21°08'S	175°12'W	Nuku'alofa, Tonga
26°20'N	127°47'E	Okinawa
17°32'S	149°34'W	Papeete, Tahiti
18°08'S	178°25'E	Suva, Fiji

Special Instructions Objective: Biological diversity
Students should choose several of the sites and find out about the natural environment, climate, and flora and fauna at each site. They should be able to discuss the reasons for the virility of certain species at each site. They should map out an itinerary for a trip to study ecosystems of the region.

Teacher's Guide
World Regions 2-8

Site: Antarctica
Assignment Focus: The Antarctica Treaty
Students should find out about the treaty—what it says and how it deals with multinational involvement. They should also be able to discuss early exploration of the region, as well as the development of the treaty and its history.

90°00'S	0°00'	Amundsen-Scott, United States
66°17'S	110°32'E	Casey, Australia
66°40'S	140°01'E	Dumont d'Urville, France
75°31'S	26°38'W	Halley Bay, United Kingdom
77°51'S	166°37'E	McMurdo, United States
70°46'S	11°50'E	Novolazerevskaja, Russia
60°40'S	44°30'W	Orcades, Argentina
62°12'S	58°55'W	Presidente Frei, Chile
70°18'S	2°22'W	Sanae, South Africa
77°51'N	166°46'E	Scott Base, New Zealand
69°00'S	39°35'E	Syowa, Japan
78°28'S	106°48'E	Vostok, Russia

Special Instructions Objective: Harsh environment
Students should report on how people have survived in such a harsh environment, from the time of the early explorers to the present. They should be able to discuss specific recorded events.

Teacher's Guide
World Regions 2-9

Site: Middle East
Assignment Focus: Politics of oil
Students should investigate the various governments' policies on the use of their oil reserves. They should be able to tell which governments are friendly to the United States, which are hostile, and which are ambivalent or undecided.

36°12'N	37°10'E	Aleppo, Syria
31°57'N	35°56'E	Amman, Jordan
33°21'N	44°23'E	Baghdad, Iraq
33°53'N	35°30'E	Beirut, Lebanon
33°30'N	36°15'E	Damascus, Syria
31°30'N	34°28'E	Ghazzah, Israel
32°33'N	35°51'E	Irbid, Jordan
31°46'N	35°14'E	Jerusalem, Israel
36°20'N	43°08'E	Mosul, Iraq
34°26'N	35°51'E	Tripoli, Lebanon

Special Instructions Objective: Comparative religions
Students should find out about the sacred holidays in Islam, Christianity, and Judaism, and report on them. They should be able to tell when the holidays fall and what they mean to the people celebrating them. Distinctive customs for some of the holidays should be included in the report.

Teacher's Guide
World Regions 2-10

Site: Eastern Europe/Western Asia
Assignment Focus: Natural vs political borders
Students should be able to determine whether the borders of the countries that are on the itinerary are natural or political borders. Students should then be able to tell from historical research which type is more effective.

33°30'N	36°15'E	Damascus, Syria
58°23'N	26°43'E	Tartu, Estonia
41°01'N	28°58'E	Istanbul, Turkey
50°26'N	30°31'E	Kiev, Ukraine
53°45'N	27°34'E	Minsk, Belarus
55°45'N	37°35'E	Moscow, Russia
43°50'N	18°25'E	Sarajevo, Bosnia
44°36'N	33°32'E	Sevastopol, Ukraine
41°43'N	44°49'E	Tbilisi, Georgia
52°15'N	21°00'E	Warsaw, Poland

Special Instructions Objective: Comparative religions
In many isolated places in this region there are tiny cultures that depend on their ability to survive in the environment. Students should be able to find four of them and report on their social structures and how they have survived, as well as how they are getting along as Western culture moves in on them.

Teacher's Guide
World Regions 2-11

Site: Slavic Europe
Assignment Focus: New governments
Students should catalogue the types of governments in this region and then look for evidence of the level of confidence the people have in their governments. Factional differences should be taken into consideration and the students should be able to predict an outcome for that country.

48°09'N	17°07'E	Bratislava, Slovakia
49°12'N	16°37'E	Brno, Czech Republic
47°30'N	19°05'E	Budapest, Hungary
47°41'N	17°38'E	Gyîr, Hungary
48°43'N	21°15'E	Kosice, Slovakia
50°03'N	19°58'E	Krakow, Poland
46°33'N	15°39'E	Maribor, Slovakia
49°45'N	13°24'E	Plzen, Czech Republic
50°05'N	14°26'E	Prague, Czech Republic

Special Instructions Objective: Cultural patterns
Students will have to search to find evidence of local cultures in the formerly Soviet countries, but the search will be rewarding and interesting. They should also report on what suppression has done to the local cultures. Where has suppression strengthened the local culture and where has it extinguished it?

Teacher's Guide
World Regions 2-12

Site: North Polar Region
Assignment Focus: International use of the North Pole
Students should find out how all the nations that have north polar territory feel about the use of the land near the pole. They should find out about other nations that have used the region and explain those nations' interests, as well.

82°30'N	62°00'W	Alert, Canada
69°39'N	162°20'E	Ambarcik, Russia
71°17'N	156°47'W	Barrow, Alaska
69°20'N	53°35'W	Godhawn, Greenland
78°50'N	103°30'W	Isachsen, Canada
71°00'N	8°30'W	Jan Mayen Island, Norway
78°13'N	15°38'E	Longyearbyen, Norway
76°15'N	119°30'W	Mould Bay, Canada
68°58'N	33°05'E	Murmansk, Finland
69°42'N	170°17'E	Pevek, Russia
77°35'N	69°40'W	Thule, Greenland
71°36'N	128°48'E	Tiksi, Russia

Special Instructions Objective: Effects of low sunlight
Students should find out about the effects on the population of low sunlight in winter and all-day sunlight in summer. They should find out how it affects the health and daily lives of the people of the region.

Teacher's Guide
World Regions 2-13

Site: Atlantic Europe
Assignment Focus: Glacial topography

Students should find out about the natural environment of Scandinavia, especially the glaciers and the landforms that they created.

57°10'N	2°01'W	Aberdeen, United Kingdom
52°22'N	4°54'E	Amsterdam, Netherlands
50°38'N	5°34'E	Antwerp, Belgium
60°23'N	5°20'E	Bergen, Norway
44°50'N	0°34'W	Bordeaux, France
48°42'N	4°29'W	Brest, France
55°28'N	8°27'E	Esbjerg, Denmark
53°16'N	9°03'W	Galway, Ireland
43°32'N	5°40'W	Gijón, Spain
50°23'N	4°10'W	Plymouth, United Kingdom

Special Instructions Objective: Standard of living

Students should investigate and report on the factors that facilitate the high standard of living in Scandinavia. They should be able to speculate about the relationships among the causes and effects. They should also be able to make suggestions regarding how to transfer certain factors to other cultures.

Teacher's Guide
World Regions 2-14

Site: Mediterranean Europe
Assignment Focus: Early governments

Students should research the philosophies and implementation of the ancient Roman and Greek forms of government and how they affected the way of life in this region during the times they governed. Students could also report on other types of government that have held power in this region.

41°23'N	2°11'E	Barcelona, Spain
37°36'N	0°59'W	Cartegena, Spain
37°30'N	15°06'E	Catania, Italy
44°25'N	8°57'E	Genoa, Italy
35°20'N	25°08'E	Iráklion, Crete
39°27'N	2°35'E	Palma, Majorca
38°15'N	21°44'E	Patrai, Greece
42°41'N	2°53'E	Perpignan, France
40°41'N	14°47'E	Salerno, Italy
40°28'N	17°14'E	Taranto, Italy
43°07'N	5°56'E	Toulon, France
35°45'N	14°31'E	Valleta, Malta

Special Instructions Objective: Languages

Students should find out about modern languages spoken in the Balkan region and relate the languages to their roots. They should report on how the languages have been standardized and also about any regional dialects that are prominent.

Teacher's Guide
World Regions 2-15

Site: Caucasus
Assignment Focus: Domestic animals

Students should find out what domestic animals are used in the countries of the region around the Caucasus Mountains. They should list the animals by use, include information on how common or rare they are and how hardy the stock is, and decide if the animals are rare enough to warrant preservation.

40°23'N	49°51'E	Baku, Azerbaidzhan
41°38'N	41°38'E	Batumi, Georgia
39°55'N	41°14'E	Erzurum, Turkey
43°20'N	45°42'E	Grozny, Russia
40°11'N	44°30'E	Jerevan, Armenia
40°40'N	46°22'E	Kirovabad, Azerbaidzhan
45°02'N	39°00'E	Krasnodar, Russia
38°05'N	46°18'E	Tabrāz, Iran
41°43'N	44°49'E	Tbilisi, Georgia
40°59'N	39°43'E	Trabzon, Turkey

Special Instructions Objective: Life expectancy

Students should find statistics for the region about male and female life expectancy. They should compare this expectancy to that in other regions of the world, and they should learn enough about the ways of life in these places to pinpoint factors that play a role in life expectancies.

Teacher's Guide
World Regions 2-16

Site: West Equatorial Africa
Assignment Focus: Western influence

Students should research how Westernized this region of the world is. They should be able to report the differences between Western influence on cities and that on the rural areas of these countries. Students should also report on the history of Westernization in this region.

5°33'N	0°13'W	Accra, Ghana
1°51'N	9°45'E	Bata, Equatorial Guinea
7°41'N	5°02'W	Bouaké, Ivory Coast
4°03'N	9°42'E	Douala, Cameroon
11°00'N	7°30'E	Kaduna, Nigeria
6°30'N	3°30E	Lagos, Nigeria
0°23'N	9°27'E	Libreville, Gabon
6°19'N	10°48'W	Monrovia, Liberia
9°24'N	0°50'W	Tamale, Ghana
3°52'N	11°31'E	Yaoundé, Cameroon

Special Instructions Objective: Role of women

Students should look into and report on the role of women in the societies of these regions. They should also be able to tell about the difference in attitudes toward women between the people in the cities and those in the rural areas.

Teacher's Guide
World Regions 2-17

Site: East Equatorial Africa
Assignment Focus: Protected species

Students should find out about the kinds and numbers of animals that are protected by the governments that have territory in the huge Serengeti Plain. They should also report on the food web and whether it is affected by the protection of the species in the region.

3°23'S	29°22'E	Bujumbura, Burundi
6°11'S	35°45'E	Dodoma, Tanzania
0°19'N	32°35'E	Kampala, Uganda
1°57'S	30°04'E	Kigali, Rwanda
2°03'N	45°22'E	Mogadishu, Somalia
4°03'S	39°40'E	Mombasa, Kenya
2°30'S	32°54'E	Mwanza, Tanzania
1°17'S	36°49'E	Nairobi, Kenya
5°04'S	39°06'E	Tanga, Tanzania
6°10'S	39°11'E	Zanzibar, Zanzibar

Special Instructions Objective: Masai way of life

Students should investigate and report on the way of life of the Masai people. All facets of their daily life should be taken into account, especially the ways in which they meet their essential needs for food, clothing, and shelter.

Teacher's Guide
World Regions 2-18

Site: Arabian Peninsula
Assignment Focus: Water resources, technology

Students should research the ways that the people of this region get and use fresh water. They should report on the ways in which they use new and developing technology to make the process easier and more efficient.

24°28'N	54°22'E	Abu Dhabi, UAE
12°46'N	45°01'E	Aden, Yemen
14°48'N	42°57'E	Al Hudaydah, Yemen
29°20'N	47°59'E	Al Kuwayti, Kuwait
25°17'N	51°32'E	Doha, Qatar
23°29'N	58°33'E	Masqat, Oman
21°27'N	39°49'E	Mecca, Saudi Arabia
24°28'N	39°36'E	Medina, Saudi Arabia
24°38'N	46°43'E	Riyadh, Saudi Arabia
28°23'N	36°35'E	Tabuk, Saudi Arabia

Special Instructions Objective: Wealth from oil

Students should find out about how the people of this region, who follow a basically nomadic lifestyle, use their wealth from the sale of the petroleum reserves in the region. Students should look for individual accounts of people using their wealth as well as reports about how the cities and governments use the wealth.

Teacher's Guide
World Regions 2-19

Site: Iranian Plateau
Assignment Focus: History, foreign domination

Students should find out how the countries of the region have been controlled by foreign governments for much of their recent history. They should report on any remaining influences in the region, even though the countries are independent now.

30°10'N	48°50'E	Abadan, Iran
37°57'N	59°23'E	Ashabad, Turmenistan
34°20'N	62°12'E	Herat, Afganistan
25°38'N	57°46'E	Jask, Iran
34°30'N	69°00'E	Kabul, Afganistan
30°17'N	57°05'E	Kerman, Iran
37°36'N	61°50'E	Mary, Turkmenistan
36°18'N	59°36'E	Mashad, Iran
30°12'N	67°00'E	Quetta, Pakistan
34°48'N	48°30'E	Hamadan, Iran

Special Instructions Objective: Women and families

Students should investigate the role that is assigned to women in this culture and how the family is structured in a traditional setting. They should be able to point out how the roles are changing and where the changes are coming from. They should also find out about the sources of the attitudes toward women.

Teacher's Guide
World Regions 2-20

Site: Central Asia
Assignment Focus: Population density

Students should do research about the population centers of Central Asia. They should be able to identify the factors that make a place attract dense populations and another not. They should also report on how some populations have adapted to climatic or environmental factors.

43°15'N	76°57'E	Alma Ata, Kazakhstan
38°35'N	68°48'E	Dusanbe, Tajikistan
37°07'N	79°55'E	Hotan, China
49°50'N	73°10'E	Karaganda, Kazakhstan
39°29'N	75°58'E	Kashi, China
45°50'N	62°10'E	Novokasalinsk, Kazakhstan
39°40'N	66°58'E	Samarkand, Uzbekistan
50°28'N	80°13'E	Semipalatinsk, Kazakhstan
41°20'N	69°18'E	Taskent, Uzbekistan
43°48'N	87°35'E	Örumqi, China

Special Instructions Objective: Climate/Environment

Students should fictionalize four fellow members, give them personalities and distinctive physiologies, and make recommendations for placing them in climatic and environmental regions that would suit their particular needs.

Teacher's Guide
World Regions 2-21

Site: Southeast Asia
Assignment Focus: Forest products

Students should find out about the harvest of mahogany, teak, rubber, and other forest products. They should be able to determine whether the governments have a plan for long-term use by how they manage forest practices. They should also report on the monsoon and how it affects the forest industry.

13°45'N	100°31'E	Bangkok, Thailand
19°45'N	99°50'E	Chiang Rai, Thailand
16°04'N	108°13'E	DaNang, VietNam
23°43'N	90°25'E	Dhaka, Bangledesh
21°02'N	105°51'E	Hanoi, VietNam
10°45'N	106°40'E	Ho Chi Minh City, VietNam
22°00'N	96°05'E	Mandalay, Burma
12°26'N	98°36'E	Mergui, Burma
11°33'N	104°55'E	Phnom Penh, Cambodia
16°47'N	96°10'E	Rangoon, Burma
20°09'N	92°54'E	Sittwe, Burma
17°58'N	102°36'E	Viangchán, Laos

Special Instructions Objective: History, colonialism

Students should investigate and report on the many national governments that have controlled most of the countries in Southeast Asia. They should report on the history of that control and determine if there is still influence from the colonial period in those sites.

Teacher's Guide
World Regions 2-22

Site: The Orient
Assignment Focus: Food, marine products

Students should research the fishing and aquaculture industries and report on the local people's diets related to those industries. Students should assess the resource management policies and determine if the focus is long range or short range.

43°51'N	125°20'E	Changchun, China
30°45'N	104°04'E	Chengdu, China
28°55'N	121°39'E	Dalian, China
20°52'N	106°41'E	Hai Phong, VietNam
42°45'N	140°43'E	Hakodate, Japan
31°36'N	130°33'E	Kagoshima, Japan
34°41'N	135°10'E	Kobe, Japan
14°35'N	121°00'E	Manila, Philippines
35°06'N	129°03'E	Pusan, Korea
36°05'N	120°21'E	Qingdao, China
25°03'N	121°30'E	Taipei, Taiwan
30°30'N	114°20'E	Wuhan, China

Special Instructions Objective: History, ancient culture

Students should find out about the high level of advancement in the cultures of the East in their early history. They should give examples and compare those cultures with the current society and cultures of that region. They should cite factors that caused a decline of the culture and when it took place.

Teacher's Guide
World Regions 2-23

Site: Romania, Bulgaria
Assignment Focus: Industry

Students should investigate the types of industry in the region and report on what they produce, how efficient they are, and how they affect the environment. They should determine where the industries get their raw materials and how the plants are powered.

47°45'N	26°40'E	Botosani, Romania
44°26'N	26°06'E	Bucharest, Romania
44°11'N	28°39'E	Constanta, Romania
44°19'N	23°48'E	Craiova, Romania
45°27'N	28°03'E	Galati, Romania
47°10'N	27°36'E	Iasi, Romania
42°09'N	24°45'E	Plovdiv, Bulgaria
43°50'N	25°57'E	Ruse, Bulgaria
45°48'N	24°09'E	Sibiu, Romania
42°43'N	23°25'E	Sofia, Bulgaria
45°45'N	21°13'E	Timisoara, Romania
43°13'N	27°55'E	Varna, Bulgaria

Special Instructions Objective: Recreation sites

Students should report on the recreation industry in this region. What types of recreation are offered, who uses them, where are they, and what do they offer compared to resorts in other parts of the world? Students should be able to recommend new sites or improvements to the present ones.

Teacher's Guide
World Regions 2-24

Site: Ukraine, Moldova, Belarus
Assignment Focus: History, strategic areas

Students should find out about the countries that have controlled this region historically and about the parts of the region that they thought were strategic. The Crimea is a prime example.

52°06'N	23°42'E	Brest, Belarus
51°30'N	31°18'E	Cernigov, Ukraine
48°00'N	37°48'E	Doneck, Ukraine
52°25'N	31°00'E	Gomel, Belarus
50°00'N	36°15'E	Harkov, Ukraine
50°26'N	30°31'E	Kiev, Ukraine
46°59'N	28°52'E	Kisinev, Moldava
53°45'N	27°34'E	Minsk, Belarus
47°33'N	30°41'E	Nicolajevka, Ukraine
46°28'N	30°44'E	Odessa, Ukraine
46°50'N	29°37'E	Tiraspol, Moldava
55°12'N	30°11'E	Vitebsk, Belarus

Special Instructions Objective: Use of rivers

Students should look at how the people make a living in this region and what part the many large rivers that cross the region play in their daily lives. Students could chart the uses of the rivers to show the diverse uses.

Teacher's Guide
World Regions 2-25

Site: Estonia, Latvia, Lithuania
Assignment Focus: Industrialization

Students should find out about types of industries and about the strength of the industrial-based economy in this region. They should find out why these areas are industrialized, when industrialization occurred, where the sources of raw materials are, and so on.

56°39'N	23°41'E	Jelgava, Latvia
63°45'N	23°54'E	Kannus, Lithuania
55°43'N	21°07'E	Klaipeda, Lithuania
56°35'N	21°01'E	Liepaja, Latvia
59°23'N	28°11'E	Narva, Estonia
58°42'N	24°32'E	Pärnu, Estonia
56°57'N	24°06'E	Riga, Latvia
59°25'N	24°45'E	Tallinn, Estonia
58°23'N	26°45'E	Tartu, Estonia
54°41'N	25°19'E	Vilnius, Lithuania

Special Instructions Objective: Germanic/Russian history

Students should investigate and report on the culture of the people that originally populated the region. They should report on the conquest by the Germanic people in the thirteenth century and about Russian domination in the twentieth century. They should also report on the influence of those peoples and how much it changed the original culture.

Teacher's Guide
World Regions 2-26

Site: Island Nations
Assignment Focus: Forms of government

Students should find out about each island nation's form of government. They should be able to tell what kind it is, how stable it is, who is in charge, what other countries each is aligned with, and what kind of world influence it has.

26°00'N	50°29'E	Bahrain
16°00'N	24°00'W	Cape Verde
21°30'N	80°00'W	Cuba
35°00'N	33°00'E	Cyprus
19°00'N	72°25'W	Haiti
65°00'N	18°00'W	Iceland
53°00'N	8°00'W	Ireland
38°00'N	137°00'E	Japan
19°00'S	46°00'E	Madagascar
35°50'N	14°30'E	Malta
5°00'S	140°00'E	New Guinea
41°00'S	174°00'E	New Zealand
13°00'N	122°00'E	Philippines
7°40'N	80°50'E	Sri Lanka

Special Instructions Objective: Unique animals/plants

Students should find out about the effect of species isolation on each of the islands in the group. They should be able to find specific examples of unique animal and plant life that are suited to the ecosystem.

Teacher's Guide
World Regions 2-27

Site: Landlocked Nations
Assignment Focus: Relations with neighbors

Students should find out how the countries that are visited get along with their neighbors. They should learn about border crossings, import/export agreements, and other governmental functions that might be important to a landlocked nation.

33°00'N	65°00'E	Afghanistan
47°30'N	14°00'E	Austria
53°50'N	28°00'E	Belarus
17°00'S	65°00'W	Bolivia
22°00'S	24°00'E	Botswana
15°00'N	19°00'E	Chad
47°00'N	20°00'E	Hungary
31°00'N	36°00'E	Jordan
18°00'N	105°00'E	Laos
47°00'N	29°00'E	Moldava
47°00'N	104°00'E	Mongolia
23°00'S	58°00'W	Paraguay
46°00'N	8°30'E	Switzerland
1°00'N	32°00'E	Uganda

Special Instructions Objective: Cultural influence

Students should investigate these landlocked countries for each one's cultural identity. They should be able to tell how little or how much that identity is influenced by the cultures of neighboring countries.

Teacher's Guide
World Regions 2-28

Site: Countries with Many Neighbors
Assignment Focus: Relations with neighbors

Students should find out how these countries get along with their neighbors. They should learn about border crossings and alignments and other agreements that might be important to a nation with many neighbors.

15°00'N	30°00'E	Sudan
35°00'N	105°00'E	China
39°00'N	35°00'E	Turkey
46°00'N	2°00'E	France
51°00'N	9°00'E	Germany
6°00'S	35°00'E	Tanzania
6°30'S	13°30'E	Zaire
60°00'N	100°00'E	Russia
9°00'S	53°00'W	Brazil

Special Instructions Objective: Cultural influence

Students should investigate these many-neighbored countries for cultural strength and influence. They should be able to tell how little or how much that influence is felt or how strong the effect is on the neighbors. They should be able to cite the factors that influence how one country affects the other and why.

3

UNITED STATES REGIONS

Teacher's Guide

Teacher's Guide
United States Regions 3-1

Site: New England
Assignment Focus: Industry, marine resources
Students should be able to find material on the history of the fishing and shellfish industry in New England and evidence of its decline in recent years. Students should suggest ways of preserving the industry and justify those suggestions.

44°49'N	68°47'W	Bangor, Maine
44°29'N	71°10'W	Berlin, New Hampshire
42°21'N	71°04'W	Boston, Massachusetts
41°11'N	73°11'W	Bridgeport, Connecticut
44°28'N	73°14'W	Burlington, Vermont
46°52'N	68°01'W	Caribou, Maine
41°46'N	72°41'W	Hartford, Connecticut
43°39'N	70°17'W	Portland, Maine
43°03'N	70°47'W	Portsmouth, New Hampshire
41°50'N	71°25'W	Providence, Rhode Island
43°37'N	72°59'W	Rutland, Vermont
42°07'N	72°36'W	Springfield, Massachusetts

Special Instructions Objective: History, personal stories
Students should find five sites that have historical significance. They should also find related anecdotal information about people or events. Students could create fictional stories based on actual occurrences and facts about those places.

Teacher's Guide
United States Regions 3-2

Site: Mid-Atlantic States
Assignment Focus: Megalopolis
Students should begin by finding information about the early years of the large population centers. They should report on how these areas grew and why and predict the future of these areas given certain conditions.

42°39'N	73°45'W	Albany, New York
39°27'N	74°35'W	Atlantic City, New Jersey
39°17'N	76°37'W	Baltimore, Maryland
42°54'N	78°53'W	Buffalo, New York
38°21'N	81°38'W	Charleston, West Virginia
39°10'N	75°32'W	Dover, Delaware
40°44'N	74°11'W	Newark, New Jersey
38°40'N	76°14'W	Norfolk, Virginia
40°26'N	80°00'W	Pittsburgh, Pennsylvania
37°16'N	79°57'W	Roanoke, Virginia
41°24'N	75°40'W	Scranton, Pennsylvania
40°05'N	80°43'W	Wheeling, West Virginia

Special Instructions Objective: Way of life, melting pot
Students should report on the several waves of immigration near the turn of the century. They should be able to find out where each group came from and what the people did when they got here. Students should be able to organize a presentation.

Teacher's Guide
United States Regions 3-3

Site: Southeast States
Assignment Focus: Resorts, historical/recreational
Students should find information on the leading tourist attractions in the Southeast with historical interest, active recreational sites, and leisure sites. They should chart them and identify them according to what they offer.

30°23'N	91°11'W	Baton Rouge, Louisiana
33°31'N	86°49'W	Birmingham, Alabama
32°48'N	79°57'W	Charleston, South Carolina
35°59'N	78°54'W	Durham, North Carolina
32°18'N	90°12'W	Jackson, Mississippi
35°58'N	83°56'W	Knoxville, Tennessee
35°08'N	90°03'W	Memphis, Tennessee
25°46'N	80°12'W	Miami, Florida
28°32'N	81°23'W	Orlando, Florida
30°25'N	87°13'W	Pensacola, Florida
32°04'N	81°05'W	Savannah, Georgia
32°30'N	93°45'W	Shreveport, Louisiana

Special Instructions Objective: Way of life, swampland
Students should be able to find information on the major wetland regions of the Southeast. They should report on the history of settlement in these places and tell how people live in these regions today. They should profile family life in this region using true or fictional accounts based on fact.

Teacher's Guide
United States Regions 3-4

Site: Great Lakes States
Assignment Focus: Oil, coal, and iron ore production
Students should locate the major mineral fields in this region and report on how productive they are, the standards of mining that are followed, the environmental posture of the producers, and the future of the resource as a viable industry.

37°00'N	86°27'W	Bowling Green, Kentucky
39°06'N	84°31'W	Cincinnati, Ohio
39°51'N	89°32'W	Decatur, Illinois
45°00'N	87°30'W	Green Bay, Wisconsin
38°03'N	84°30'W	Lexington, Kentucky
42°43'N	87°48'W	Racine, Wisconsin
42°17'N	89°06'W	Rockford, Illinois
43°25'N	83°58'W	Saginaw, Michigan
46°30'N	84°21'W	Sault Ste. Marie, Michigan
48°00'N	88°00'W	Superior, Wisconsin
39°28'N	87°24'W	Terre Haute, Indiana
41°39'N	83°32'W	Toledo, Ohio

Special Instructions Objective: Way of life, the miner
Students should find out what life is like in a coal, iron ore, or oil mining town. They should find out how people live and how the industry affects their daily lives.

Teacher's Guide
United States Regions 3-5

Site: Plains States
Assignment Focus: Shelter, residential
Students should contact businesses and chambers of commerce in the plains states and find out what people are using for residential building materials. They should find out about the history of architecture in this region and determine how it has been affected by the local environment.

41°59'N	91°40'W	Cedar Rapids, Iowa
37°45'N	100°00'W	Dodge City, Kansas
46°52'N	96°48'W	Fargo, North Dakota
37°06'N	94°31'W	Joplin, Missouri
34°44'N	92°15'W	Little Rock, Arkansas
44°22'N	100°21'W	Pierre, South Dakota
33°30'N	111°56'W	Scottsdale, Nebraska
42°30'N	96°23'W	Sioux City, Nebraska
45°33'N	94°10'W	St. Cloud, Minnesota
36°09'N	95°58'W	Tulsa, Oklahoma
48°09'N	103°37'W	Williston, North Dakota

Special Instructions Objective: Shelter, history
Students should research how the early pioneers that settled the region met their needs for shelter. The report should include information on the sod house. They should be able to tell how people lived and how they withstood the elements.

Teacher's Guide
United States Regions 3-6

Site: Rocky Mountain States
Assignment Focus: Government, public land
Students should make a catalogue of public lands in this region. They should also find out what kinds of activity are allowed on these properties and make judgments as to whether the activities are good for the lands.

40°01'N	105°17'W	Boulder, Colorado
42°51'N	106°19'W	Casper, Wyoming
39°05'N	108°33'W	Grand Junction, Colorado
47°30'N	111°17'W	Great Falls, Montana
41°19'N	105°35'W	Laramie, Wyoming
46°25'N	117°01'W	Lewiston, Idaho
41°44'N	111°50'W	Logan, Utah
46°52'N	114°01'W	Missoula, Montana
40°14'N	111°39'W	Provo, Utah
42°34'N	114°28'W	Twin Falls, Idaho

Special Instructions Objective: Cattle ranching
Students should find out about cattle ranching and compare the modern rancher to the stereotype of the rugged cattle rancher. They should find out about the lives of ranchers and about the factors that are involved in this region's reputation of having a very high per capita income and a high standard of living.

Teacher's Guide
United States Regions 3-7

Site: Southwest States
Assignment Focus: National parks
Students should survey the national parks. They should discuss the history of the parks, their exploration, and their acceptance into the park system. They should recommend getting a fictional site accepted into the system.

35°05'N	106°40'W	Albuquerque, New Mexico
35°13'N	101°46'W	Amarillo, Texas
39°10'N	119°46'W	Carson City, Nevada
34°24'N	103°12'W	Clovis, New Mexico
37°16'N	107°53'W	Durango, Colorado
31°45'N	106°29'W	El Paso, Texas
39°15'N	114°53'W	Ely, Nevada
35°12'N	111°39'W	Flagstaff, Arizona
27°31'N	99°30'W	Laredo, Texas
33°35'N	101°51'W	Lubbock, Texas
32°13'N	110°58'W	Tucson, Arizona
31°55'N	97°08'W	Waco, Texas
32°43'N	114°37'W	Yuma, Arizona

Special Instructions Objective: Border with Mexico
Students should report on U.S.–Mexico relations concerning the border and immigration. They should find reasons why the problem is so intense and make suggestions about improvement. They should discuss the effects of NAFTA and what it will mean to the region.

Teacher's Guide
United States Regions 3-8

Site: Pacific States
Assignment Focus: Natural disasters
Students should find information about natural disasters in the last hundred years. They should find out how the local people coped. Students should identify the factors that would make a person choose to live in such an insecure area.

46°11'N	123°50'W	Astoria, Oregon
35°23'N	119°01'W	Bakersfield, California
48°46'N	122°29'W	Bellingham, Washington
44°02'N	123°05'W	Eugene, Oregon
40°47'N	124°09'W	Eureka, California
36°45'N	119°45'W	Fresno, California
42°13'N	121°46'W	Klamath Falls, Oregon
32°43'N	117°09'W	San Diego, California
35°17'N	120°40'W	San Luis Obispo, California
47°40'N	117°23'W	Spokane, Washington
37°57'N	121°17'W	Stockton, California
46°36'N	120°31'W	Yakima, Washington

Special Instructions Objective: Local health issues
Students research the educational systems and report on their condition. They should find out about state initiatives and about any problems, and they should make a judgment about what should be done in the future.

Teacher's Guide
United States Regions 3-9

Site: Alaska
Assignment Focus: Natural resources

Students should make a catalogue or chart of the living and mineral natural resources in Alaska. They should be able to tell which resources are valuable, accessible, and plentiful.

61°13'N	149°53'W	Anchorage
71°17'N	156°47'W	Barrow
59°02'N	158°29'W	Dillingham
53°53'N	166°32'W	Dutch Harbor
64°51'N	147°43'W	Fairbanks
66°34'N	145°14'W	Fort Yukon
55°21'N	131°35'W	Ketchikan
64°30'N	165°24'W	Nome
59°28'N	185°19'W	Skagway
61°07'N	146°16'W	Valdez

Special Instructions Objective: History, settlement

Students should research the history of settlement and immigration in Alaska during the time the United States has owned it. They should be able to cite factors and forces that have shaped the population growth patterns over the years.

Teacher's Guide
United States Regions 3-10

Site: Island States and Territories
Assignment Focus: Climate and weather patterns

Students should find out about and report on climate and weather conditions. They should be able to discuss seasonal variations and should recommend a site for installation of a satellite tracking system. The prerequisites for such an installation should be part of the final report.

13°28'N	144°45'E	Agana, Guam
18°21'N	64°56'W	Charlotte Amalie, St. Thomas
17°45'N	64°40'W	Christiansted, St. Croix
19°44'N	155°05'W	Hilo, Hawai'i
21°19'N	157°52'W	Honolulu, Hawai'i
7°20'N	134°30'E	Koror, Palau Island
28°13N	177°22'W	Midway Island
14°16'S	170°42'W	Pago Pago, Amer. Samoa
18°28'N	66°07'W	San Juan, Puerto Rico
19°18'N	166°36'W	Wake Island

Special Instructions Objective: History, acquisition

Students should learn how each site came to be under the control of the United States. They should find out how the people feel about the United States and about other sites that used to be part of the group and are now independent.

4

CONTINENTAL
TOPICS

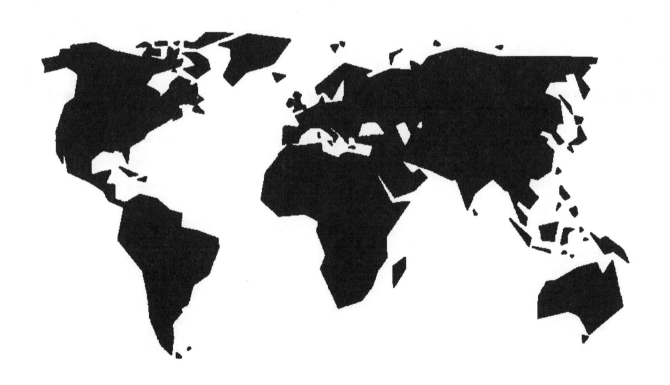

Teacher's Guide

Teacher's Guide
Continental Topics 4-1

Site: Africa
Assignment Focus: African cities

Students should make recommendations as to safe sites for retiring members' homes. They should consider such factors as ease of settlement, natural environment, language, trade with Western countries, and acceptance of Western culture.

9°01'N	38°46'E	Adis Abeba, Ethiopian plateau
4°22'N	18°35'E	Bangui, equatorial rain forest
19°50'S	34°52'E	Biera, southeast coastal
12°00'N	8°31'E	Kano, interior grassland
15°50'N	33°00'E	Khartoum, Nile River valley
0°06'S	34°45'E	Kisumu, Lake Victoria
31°38'N	8°00'W	Marrakech, Atlantic Desert
4°48'S	11°51'E	Pointe Noire, coastal rain forest
16°46'N	2°59'W	Tombouctou, sub-Sahara grassland
24°01'S	21°43'E	Tshane, Kalahari Desert

Special Instructions Objective: Geriatric pathology and environmental compatibility

Students should be looking for a list of common ailments in an aging population and should describe the natural environment and climate that would best suit someone with each ailment.

Teacher's Guide
Continental Topics 4-2

Site: African Ports
Assignment Focus: Imports and exports

Students should find out from statistical sources the types, volume, and values of imports and exports, and the percent of the world market that they represent.

5°19'N	4°02'W	Abidjan, Ivory Coast
30°03'N	31°15'E	Cairo, Egypt
35°55'S	18°22'E	Cape Town, South Africa
33°37'N	7°35'W	Casablanca, Morocco
14°40'N	17°26'W	Dakar, Senegal
6°30'N	3°30'E	Lagos, Nigeria
8°50'S	13°15'E	Luanda, Angola
26°00'S	32°30'E	Maputo, Mozambique
2°03'N	45°22'E	Mogadishu, Somalia
32°54'N	13°11'E	Tripoli, Libya

Special Instructions Objective: Conflict

Students should find out about the presence of conflict, armed or otherwise, in any of the sites. They should report on the state of the conflict and the causes and make predictions about the endurance of the conflict and its final outcome or resolution. They should also cite places that are stable and not likely to be involved in conflict.

Teacher's Guide
Continental Topics 4-3

Site: North America Port Cities
Assignment Focus: Industry

Students should find out what major industries are located in each of these major port cities and why they are there. They should find out what human and natural resources the industries have and how the industries are doing in the world market.

61°13'N	149°53'W	Anchorage, Alaska
42°21'N	71°04'W	Boston, Massachusetts
44°39'N	63°36'W	Halifax, Canada
23°08'N	82°22'W	Havana, Cuba
29°46'N	95°22'W	Houston, Texas
38°40'N	76°14'W	Norfolk, Virginia
46°49'N	71°13'W	Québec, Canada
37°48'N	122°24'W	San Francisco, California
32°04'N	81°05'W	Savannah, Georgia
47°34'N	52°43'W	St. John's, Newfoundland
49°16'N	123°07'W	Vancouver, British Columbia
19°12'N	96°08'W	Veracruz, Mexico

Special Instructions Objective: Historical attractions

Students should find out why these cities have become major centers for business, industry, and commerce. They should look for historical data and attractions that would contribute to the growth of the site.

Teacher's Guide
Continental Topics 4-4

Site: North America
Assignment Focus: Elevation

Students should list the elevation of each city and then determine if there are any good or bad effects on people working at certain jobs or careers in that region because of the elevation.

39°06'N	84°31'W	Cincinnati, Ohio
64°04'N	139°25'W	Dawson, Canada
39°05'N	108°33'W	Grand Junction, Colorado
20°40'N	103°20'W	Guadalajara, Mexico
35°58'N	83°56'W	Knoxville, Tennessee
48°14'N	101°18'W	Minot, North Dakota
46°06'N	64°07'W	Moncton, Canada
25°40'N	100°19'W	Monterrey, Mexico
28°32'N	81°23'W	Orlando, Florida
52°07'N	106°38'W	Saskatoon, Saskatchewan
47°40'N	117°23'W	Spokane, Washington
46°30'N	81°00'W	Sudbury, Ontario, Canada

Special Instructions Objective: Professional sports

Students should find out about any professional sports teams that are from these sites and how successful they are in their leagues. Students should determine how the team reflects the spirit and the character of the people of that city.

Teacher's Guide
Continental Topics 4-5

Site: South America 1
Assignment Focus: Environmental site match
Students should determine the character of the environment, climate, and topography at the sites, then suggest a matching site in the United States. They should be able to justify their choices.

23°39'S	70°42'W	Antofagasto, Chile
23°16'S	57°40'W	Asunción, Paraguay
3°27'N	76°31'W	Cali, Colombia
20°27'S	60°50'W	Campo Grande, Brazil
13°31'S	71°59'W	Cusco, Peru
2°10'S	79°50'W	Guayaquil, Ecuador
9°45'N	63°11'W	Maturín, Venezuela
32°54'S	68°50'W	Mendoza, Argentina
34°53'S	56°11'W	Montevideo, Uruguay
17°48'S	63°10'W	Santa Cruz, Bolivia

Special Instructions Objective: Local celebrations and music
Students should find out about local celebrations. They should give reasons for the celebrations, describe the music that is associated with them, and tell about the instruments that are used to make the music. Students could draw the instruments and perhaps even reproduce the music.

Teacher's Guide
Continental Topics 4-6

Site: South America 2
Assignment Focus: Environmental site match
Students should determine the character of the environment, climate, and topography at the sites that they are visiting and then suggest a matching site in Europe or Asia. They should be able to justify their choices.

1°15'S	78°37'W	Ambato, Ecuador
10°59'N	74°48'W	Barranquilla, Colombia
8°17'S	35°58'W	Caruaru, Brazil
23°25'S	51°17'W	Concepción, Paraguay
20°13'S	70°10'W	Iquique, Chile
3°50'S	73°15'W	Iquitos, Peru
19°35'S	65°45'W	Potosi, Bolivia
32°57'S	60°40'W	Rosario, Argentina
31°23'S	57°58'W	Salto, Uruguay
10°11'N	67°45'W	Valencia, Venezuela

Special Instructions Objective: Local health issues
Students should find out about family life in or near each of the sites. They should be able to tell how people in each of the regions meet their needs for food, what they like to eat, and how they prepare it. Students could try making certain foods for tasting.

Teacher's Guide
Continental Topics 4-7

Site: Europe 1
Assignment Focus: Government, monarchy
Students should find out about the history of the monarchies in the countries listed. They should find out how each monarchy worked and how it ended or came to the status that it is in today.

57°03'N	9°56'E	Elborg, Denmark
53°20'N	6°15'W	Dublin, Ireland
48°43'N	21°15'E	Kosice, Slovakia
51°24'N	16°13'E	Lubin, Poland
36°43'N	4°25'W	Málaga, Spain
48°41'N	6°12'E	Nancy, France
38°07'N	13°22'E	Palermo, Italy
48°46'N	9°11'E	Stuttgart, Germany
40°38'N	22°56'E	Thessaloniki, Greece
45°48'N	16°00'E	Zagreb, Croatia

Special Instructions Objective: Languages
Students should catalogue the languages that are spoken in the regions around the sites and find out about the roots of the languages in their spoken and written forms.

Teacher's Guide
Continental Topics 4-8

Site: Europe 2
Assignment Focus: Government alliances
Students should be able to find information on all of the important organizations to which the governments of Europe belong. They should be able to find out about the purposes of the organizations, who belongs to them, and how they have grown and changed over time.

44°50'N	20°30'E	Belgrade, Yugoslavia
52°31'N	13°24'E	Berlin, Germany
37°53'N	4°46'W	Córdoba, Spain
55°53'N	4°15'W	Glasgow, United Kingdom
38°43'N	9°08'W	Lisbon, Portugal
48°52'N	2°20'E	Paris, France
42°45'N	12°29'E	Rome, Italy
59°20'N	18°03'E	Stockholm, Sweden
48°12'N	16°22'E	Vienna, Austria
47°20'N	8°35'E	Zurich, Switzerland

Special Instructions Objective: Products
Students should be able to choose a product that is made or produced in each site that characterizes the region and the people of that region. They should produce an attractive display of the products and they should justify their choices.

Teacher's Guide
Continental Topics 4-9

Site: Central America and the Caribbean 1
Assignment Focus: Spanish domination and the slave trade

Students should research the history of the region and find out about the slave trade and Spanish conquest. They should describe continuing influences related to these factors—what effects have they had and how are these manifested?

13°06'N	59°37'W	Bridgetown, Barbados
9°22'N	79°54'W	Colón, Panama
15°47'N	86°50'W	La Ceiba, Honduras
12°35'N	86°35'W	León, Nicaragua
18°30'N	77°55'W	Montego Bay, Jamaica
15°43'N	88°36'W	Puerto Barrios, Guatemala
16°07'N	88°48'W	Punta Gorda, Belize
9°58'N	84°50'W	Puntarenas, Costa Rica
22°24'N	79°58'W	Santa Clara, Cuba
19°27'N	70°42'W	Santiago, Dom. Rep.
17°06'N	61°51'W	St. John's, Antigua
12°06'N	68°56'W	Willemstad, Curaçao

Special Instructions Objective: Local health issues

Students should look at the marine or forest activity of each site and determine what benefits the population is gaining. Students could catalogue the information in an attractive format for presentation.

Teacher's Guide
Continental Topics 4-10

Site: Central America and the Caribbean 2
Assignment Focus: Regional conflict

Students should find out which of the governments in the region have been under foreign control and how the countries fared when they became independent. They should cite the nations that are involved in conflict, talk about the nature of that conflict, and predict its outcomes.

17°15'N	88°46'W	Belmopan, Belize
19°45'N	72°15'W	Cap-Haätien, Haäti
8°25'N	82°27'W	David, Panama
18°00'N	76°50'W	Kingston, Jamaica
10°00'N	83°02'W	Limón, Costa Rica
12°53'N	85°57'W	Matagalpa, Nicaragua
16°14'N	61°23'W	Pointe-à-Pitre, Guadeloupe
18°01'N	66°37'W	Ponce, Puerto Rico
14°50'N	91°31'W	Quezaltenango, Guatemala
13°29'N	88°11'W	San Miguel, El Salvador
20°01'N	75°49'W	Santiago de Cuba, Cuba
12°03'N	61°45'W	St. Georges, Grenada

Special Instructions Objective: Relations with outsiders

Students should find out how the native populations in resort and tourist areas get along with outsiders. They should also find out if outside business interests are welcomed or discouraged.

Teacher's Guide
Continental Topics 4-11

Site: Asia 1
Assignment Focus: Standard of living

Students should assess the general standard of living in each site and note who possesses the wealth and how they maintain it. They should also find information about the natural environment and make correlations between wealth and environment. They should present their conclusions in a graphically attractive report.

30°10'N	48°50'E	Abadan, Iran
18°58'N	72°50'E	Bombay, India
22°15'N	114°10'E	Hong Kong
14°35'N	121°00'E	Manila, Philippines
35°06'N	129°03'E	Pusan, Korea
16°47'N	96°10'E	Rangoon, Burma
31°41'N	121°28'E	Shanghai, China
1°17'N	103°51'E	Singapore, Malaysia
35°40'N	139°46'E	Tokyo, Japan
43°10'N	131°56'E	Vladivostok, Russia

Special Instructions Objective: Religion

Students should find out which religion most people follow in each site. They should note whether it is different from the state religion. They should characterize the religions according to the main beliefs, worldviews, and celebrations.

Teacher's Guide
Continental Topics 4-12

Site: Asia 2
Assignment Focus: Endangered species

Students should list the endangered species in the regions. They should assess the condition of the populations and how the species are faring under the protection of the governments. They should report on the species' habitat, range, and general health.

6°10'S	106°46'E	Jakarta, Indonesia
34°31'N	69°12'E	Kabul, Afghanistan
3°10'N	101°42'E	Kuala Lumpur, Malaysia
35°00'N	135°45'E	Kyoto, Japan
28°36'N	77°12'E	New Delhi, India
11°33'N	104°55'E	Phnom Penh, Kampuchea
37°34'N	127°00'E	Seoul, Korea
35°40'N	51°26'E	Tehran, Iran
47°55'N	106°53'E	Ulanbaatar, Mongolia
30°30'N	114°20'E	Wuhan, China

Special Instructions Objective: Family, role of children

Students should find out about family structures and specifically about the role of children in the family. They should find out what it means to come of age—how and when it happens for males and females.

5

URBAN TOPICS

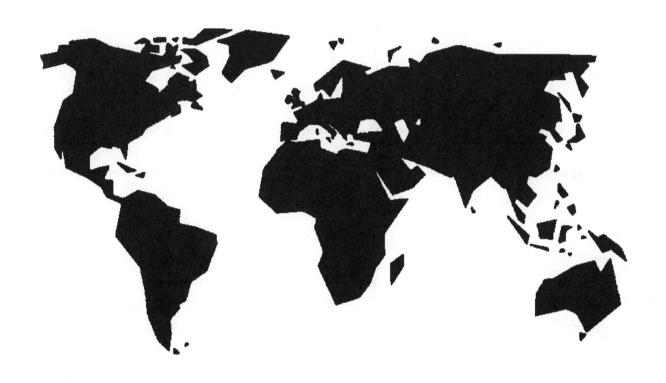

Teacher's Guide

Teacher's Guide
Urban Topics 5-1

Site: World Port Cities 1
Assignment Focus: Quality measures

Students should be looking for information about ease of access, natural harbor protection, friendly governments, world coverage, and strategic location.

12°28'S	130°50'E	Darwin, Northern Australia
44°39'N	63°36'W	Halifax, Atlantic Canada
24°10'N	110°18'W	La Paz, Baja Peninsula, Mex.
10°39'N	61°31'W	Port of Spain, Trinidad
23°39'S	70°24'W	Antafagasta, Chile
14°40'N	17°26'W	Dakar, Cape Verde, Africa
4°03'S	39°31'E	Mombasa, Kenya, Indian Ocean
51°54'N	8°28'W	Cork, southern Ireland
39°36'N	19°55'E	Kerkira, Ionian Sea, w. Greece
24°52'N	67°03'E	Karachi, Pakistan
20°52'N	106°41'E	Hai Phong, northern VietNam
41°45'N	140°43'E	Hakodate, Hokkaido Is., Japan

Special Instructions Objective: Shipbuilding ports

You should be looking for research and possibly letters to a local shipbuilder or to the federal government to gain information on shipbuilding facilities around the world.

Teacher's Guide
Urban Topics 5-2

Site: World Port Cities 2
Assignment Focus: Resources, oil

Students should research shipping at these ports and should gather statistical data. They should put the information in a chart that compares the cities according to a variety of criteria.

35°36'N	140°07'E	Chiba, Japan
29°46'N	95°22'W	Houston, Texas
35°32'N	139°43'E	Kawasaki, Japan
34°41'N	135°10'E	Kobe, Japan
43°18'N	5°24'E	Marseilles, France
29°04'N	48°09'E	Mina' al, Ahmadi
29°58'N	90°07'W	New Orleans, Louisiana
40°43'N	74°01'W	New York, New York
51°55'N	4°28'E	Rotterdam, Netherlands
35°27'N	139°39'E	Yokohama, Japan

Special Instructions Objective: The arts

Students should investigate the arts programs at these sites and find out about resident dance and music organizations and well-known museums with important holdings. Students should be able to find out about the reputation of the facilities in these sites and compare them to state-of-the-art facilities.

Teacher's Guide
Urban Topics 5-3

Site: Forbidden Cities
Assignment Focus: Local culture

Students should find out why these sites were known as forbidden cities. They should find out what the consequences were for breaking through the "forbiddenness," what these places were like at their peak, and if their influence has faded.

29°42'N	91°07'E	Lhasa, Tibet
21°27'N	39°49'E	Mecca
40°15'N	24°15'E	Mount Athos, n. Greece
11°33'N	92°15'E	N. Sentinel Is., Anduman Is.
39°55'N	116°23'W	Beijing, China
52°38'N	6°14'E	Staphorst, n.e. Amsterdam

Special Instructions Objective: Fictional account

Each student should compose a fictional account of life in the forbidden city of his or her choice. Students should include factual information about the cities without making the account sound like a report.

Teacher's Guide
Urban Topics 5-4

Site: Top 10 Large Cities
Assignment Focus: Metropolitan commonalities

Students should look at the ways of life and look for common patterns in the growth and maintenance of the life of the city. Students should report on those patterns and suggest ways to improve life.

18°58'N	72°50'E	Bombay, India
34°36'S	58°27'W	Buenos Aires, Argentina
22°32'N	88°22'E	Calcutta, India
34°03'N	118°15'W	Los Angeles, California
19°24'N	99°09'W	Mexico City, Mexico
55°45'N	37°35'E	Moscow, Russia
40°43'N	74°01'W	New York, New York
35°37'N	137°14'E	Osaka/Kobe/Kyoto, Japan
22°54'S	43°15'W	Rio de Janeiro, Brazil
23°32'S	46°37'W	São Paulo, Brazil
37°34'N	127°00'E	Seoul, Korea
35°40'N	139°46'E	Tokyo/Yokohama, Japan

Special Instructions Objective: Cultural patterns

Students look for and report on cultural patterns that have survived metropolitan life. They should determine what it is about the particular culture that enables it to survive in such a homogeneous climate. Students should make recommendations for its continued health and survival.

Teacher's Guide
Urban Topics 5-5

Site: Second 10 Large Cities
Assignment Focus: Character of a city
Students should find out about the way of life and be able to characterize each city. They should tell about the reputation of the people of that site, positive or negative, whether that reputation is deserved, and how it has changed over time.

30°03'N	31°15'E	Cairo, Egypt
28°40'N	77°13'E	Delhi, India
51°27'N	7°01'E	Essen, Germany
6°10'S	106°46'E	Jakarta, Indonesia
24°52'N	67°03'E	Karachi, Pakistan
6°27'N	3°23'E	Lagos, Nigeria
12°03'S	77°03'W	Lima, Peru
51°30'N	0°10'W	London, UK
14°35'N	121°00'E	Manila, Philippines
48°52'N	2°20'E	Paris, France
31°41'N	121°28'E	Shanghai, China
35°40'N	51°26'E	Tehran, Iran

Special Instructions Objective: Comparative religions
Students should research animals that thrive in urban and metropolitan settings. They should find out what enables each species to survive and whether the animals are involved in a food web other than one involving humans.

Teacher's Guide
Urban Topics 5-6

Site: Summer Olympic Games Cities
Assignment Focus: Politics of oil
Students should find out how Olympic host cities are chosen and what factors past host cities have in common. They should look at the criteria that the Olympic Committee uses to select a site and then recommend a site based on what they learn.

37°59'N	23°44'E	Athens, Greece
41°23'N	2°11'E	Barcelona, Spain
52°31'N	13°24'E	Berlin, Germany
60°01'N	24°58'E	Helsinki, Finland
51°30'N	0°10'W	London, UK
37°49'S	144°58'E	Melbourne, Australia
45°31'N	73°34'W	Montréal, Canada
48°09'N	11°35'E	Munich, Austria
48°52'N	2°20'E	Paris, France
42°45'N	12°29'E	Rome, Italy
37°34'N	127°00'E	Seoul, Korea
35°40'N	139°46'E	Tokyo, Japan

Special Instructions Objective: Comparative religions
Students should put the sites in chronological order based on when they hosted the Olympics and chart the outstanding athletes and innovations that characterized the Games at each site.

Teacher's Guide
Urban Topics 5-7

Site: Winter Olympic Games Cities
Assignment Focus: Environmental compatibility
Students should research facilities that have been built at past Olympic sites and determine how they fit in with the local environment. They should cite innovations in facility construction that meet the changing needs of the Olympic community.

51°03'N	114°05'W	Calgary, Alberta, Canada
45°55'N	6°52'E	Chamonix, France
46°32'N	12°08'E	Cortina d'Ampezzo, Italy
47°30'N	11°06'E	Garmisch-Partenkirchen, Ger.
45°10'N	5°43'E	Grenoble, France
47°16'N	11°24'E	Innsbruck, Austria
44°18'N	73°59'W	Lake Placid, New York
59°55'N	10°45'E	Oslo, Norway
43°03'N	141°21'E	Sapporo, Japan
43°50'N	18°25'E	Sarajevo, Yugoslavia
36°45'N	119°45'W	Squaw Valley, California
46°32'N	9°49'E	St. Moritz, Switzerland

Special Instructions Objective: Resources, snow
Students should find out how each site handled the amount of snow at the time of the Games. They should find out how the Olympic Committee selects sites for Winter Games and if the committee considers snowfall.

Teacher's Guide
Urban Topics 5-8

Site: World Capitals 1
Assignment Focus: Stability of government
Students should study governmental changes. They should determine if the changes were normal, expected, and orderly, or violent and unexpected. Students should tell about the players in any change and what they stand for.

39°56'N	32°52'E	Ankara, Turkey
40°23'N	49°51'E	Baku, Azerbaidzhan
60°01'N	24°58'E	Helsinki, Finland
6°10'S	106°46'E	Jakarta, Indonesia
34°31'N	69°12'E	Kabul, Afghanistan
15°50'N	33°00'E	Khartoum, Sudan
8°50'S	13°15'E	Luanda, Angola
6°19'N	10°48'W	Monrovia, Liberia
45°25'N	75°42'W	Ottawa, Canada
64°09'N	21°57'W	Reykjavik, Iceland
37°34'N	127°00'E	Seoul, South Korea
52°15'N	21°00'E	Warsaw, Poland

Special Instructions Objective: World leaders
Students should investigate leadership in these governments. They should profile important leaders in the government's history. Students should then give a detailed report about the leader of their choice.

Teacher's Guide
Urban Topics 5-9

Site: World Capitals 2
Assignment Focus: Government philosophy

Students should find out about various government structures, looking for not only official information but government philosophies and geopolitical alignments.

15°47'S	47°55'W	Brasilia, Brazil
50°50'N	4°20'E	Brussels, Belgium
34°36'S	58°27'W	Buenos Aires, Argentina
35°17'S	149°08'E	Canberra, Australia
23°08'N	82°22'W	Havana, Cuba
14°35'N	121°00'E	Manila, Philippines
55°45'N	37°35'E	Moscow, Russia
45°25'N	75°42'W	Ottawa, Canada
39°55'N	116°23'E	Peking, China
59°20'N	18°03'E	Stockholm, Sweden
35°40'N	51°26'E	Tehran, Iran
41°17'S	174°46'E	Wellington, New Zealand

Special Instructions Objective: Issues alignments

Look for students to find out that governments will align in various ways depending on the issue involved. Students should look at such current issues as world debt, world energy resources and distribution, deforestation, endangered species, the Antarctic, the oceans, and so on.

Teacher's Guide
Urban Topics 5-10

Site: World Capitals 3
Assignment Focus: Capitol architecture

Students should survey the architecture of the capitol buildings in the sites and determine how well it reflects the culture of the city and the nation.

9°10'N	7°11'E	Abuja, Nigeria
31°57'N	35°56'E	Amman, Jordan
48°09'N	17°07'E	Bratislava, Slovakia
13°06'N	59°37'W	Bridgetown, Barbados
53°20'N	6°15'W	Dublin, Ireland
15°25'S	28°17'E	Lusaka, Zambia
53°45'N	27°34'E	Minsk, Belarus
39°01'N	125°45'E	Pyongyang, North Korea
0°13'S	78°30'W	Quito, Equador
16°47'N	96°10'E	Rangoon, Burma
42°45'N	12°29'E	Rome, Italy
34°00'N	9°00'E	Tunis, Tunisia
38°54'N	77°01'W	Washington, D.C.

Special Instructions Objective: Historical monuments

Students should find out about any prominent monuments in the sites and describe who or what the monuments honor.

Teacher's Guide
Urban Topics 5-11

Site: Tiny Countries
Assignment Focus: Reason for existence

Students should find out what form of government each of these small countries has and how long it has been in power. They should find out why the country exists, what its purpose is, and how it has stayed independent for such a long time.

42°30'N	1°30'E	Andorra
17°03'N	61°48'W	Antigua
12°10'S	44°10'E	Comoros
12°07'N	61°40'W	Grenada
47°10'N	9°30'E	Liechtenstein
3°15'N	73°00'E	Maldives
35°50'N	14°30'E	Malta
43°42'N	7°23'E	Monaco
43°55'N	12°28'E	San Marino
1°17'N	103°51'E	Singapore

Special Instructions Objective: National character

Students should learn about the people of these small countries and find out if there is, indeed, a national character or culture. They should report on what characterizes the culture or the way people live in these places.

6
RURAL CULTURE

Teacher's Guide

Teacher's Guide
Rural Culture 6-1

Site: Western Hemisphere
Assignment Focus: Environmental/Economic
Students should find out about rural areas in this region. They should be looking for cultural and environmental data and observing the close interrelationships among environment, culture, and local economy.

60°N	155°W	Aleutian Mountains, Alaska
65°N	20°W	Iceland
40°N	100°W	Nebraska prairie
30°N	85°W	Florida panhandle swampland
30°N	115°W	Baja Peninsula
15°N	85°W	Honduras rain forest
20°N	75°W	coastal Cuba
30°N	70°W	Andes Mountains, Chile
5°S	65°W	Amazon rain forest, Brazil
45°S	60°W	Pampas grassland, Argentina

Special Instructions Objective: Economic factors

Students should be able to speculate about the effects of the gain or loss of vital segments of the local economy. Students should also consider the long-term as well as short-term effects such changes will have on the environment.

Teacher's Guide
Rural Culture 6-2

Site: Eastern Hemisphere
Assignment Focus: Cultural change
Students should find out about rural areas and research the changes that cultures in those areas have experienced over time. They should find out what factors have caused the changes and assess how the present culture is like the original primary culture.

44°S	170°E	South Island, New Zealand
20°N	106°E	North VietNam
41°N	141°E	Northern Japan, Honshu Is.
30°N	105°E	Central China
32°N	92°E	Tibet
6°N	81°E	Sri Lanka
35°N	70°E	northern Afghanistan
15°S	28°E	Zambia
36°N	6°E	northern Algeria
50°N	2°E	northwestern France
66°N	18°E	central Sweden

Special Instructions Objective: Relations with outsiders

Students should find out how people in these cultures feel about outsiders and outside influence. Students should find out how the people treat those who are not like them and determine whether this behavior correlates with the survival of the culture.

Teacher's Guide
Rural Culture 6-3

Site: World Native Cultures
Assignment Focus: Daily life
Students should write a fictional narrative that includes insight into family and group structures and relationships. They should also include information about domesticated plants and animals and their uses.

70°N	160°W	Inuit, northern Alaska
35°N	110°W	American Indian, s.w. US
20°N	90°W	Mayan culture, Yucatán, Mex.
30°S	70°W	Incan culture, Andean Indian
5°S	60°W	Arawak Indian, Amazon River Basin
5°S	35°E	Masai, Serengeti Plain, E. Africa
20°N	50°E	Arab nomads, Saudi Arabia
70°N	25°E	Sami, northern Scandinavia
18°S	178°E	Fiji Islanders, Fiji, South Pacific
45°N	105°E	Mongols, Mongolia/N. China
25°S	125°E	Aborigines, Australian outback
0°	20°E	Bantu, Congo River basin

Special Instructions Objective: Community model
Students should research a group and create a detailed drawing or model of a typical community. Students should also research the natural environment of one of the groups and create a detailed illustration of the food web.

Teacher's Guide
Rural Culture 6-4

Site: Worldwide Site Choice
Assignment Focus: Site suitability
Students should choose twelve locations—anywhere except the Antarctic—for establishing a communication network. Students should be looking for the relative proximity of the twelve locations and the political and environmental acceptability of the selected sites.

Special Instructions Objective: Communication technology

Students should be looking for innovative research into the present state of communication technology. Letters to communications providers and producers would be positive. Even speculation about the usefulness of predicted future technologies would be a useful avenue of inquiry.

7

WORLD CLIMATIC REGIONS

Teacher's Guide

Teacher's Guide
World Climatic Regions 7-1

Site: Swampland
Assignment Focus: Animals, ecosystem
Students should find out about the food webs among the different species that live in these sites. They should present the information in an attractive format.

18°50'S	62°10'W	Bañados del Izozog, Bolivia
0°00'	17°00'E	Congo Basin, Africa
2°00'N	102°30'E	Eastern Sumatra
25°52'N	81°23'W	Everglades, United States
36°30'N	76°30'W	Great Dismal Swamp
30°50'N	47°10'E	Hawr al Hammar, Iraq
37°00'N	6°15'W	Las Marismas, Spain
45°30'N	29°45'E	Mouths of the Danube, Rom.
4°50'N	6°00'E	Niger River basin
30°42'N	82°20'W	Okefenokee swamp
18°00'S	56°00'W	Pantanal de São Lorenço, Br.
19°00'S	52°00'W	Pantanal do Rio Negro, Br.
3°00'N	114°00'E	southern Borneo
5°24'N	0°20'W	The Fens, United Kingdom

Special Instructions Objective: History, land use
Students should find the history of the use of these different swamp regions. If there is folklore, students should become familiar with it and report on it.

Teacher's Guide
World Climatic Regions 7-2

Site: Rain Forest
Assignment Focus: Health, medicinal plants
Students should find out about some of the rare and common medicinal plants in the rain forests. They should find out about scientists' efforts to preserve this valuable resource and propose ideas to help solve the problem.

17°15'N	88°46'W	Belmopan, Belize
23°43'N	90°25'E	Dhaka, Bangladesh
3°43'S	38°30'W	Fortaleza, Brazil
20°52'N	106°41'E	Hai Phong, VietNam
4°22'N	7°43'W	Harper, Liberia
2°32'S	140°42'E	Jayapura, Papua New Guinea
0°25'N	25°12'E	Kisangani, Zaire
16°42'N	74°13'E	Kolhapur, India
3°10'N	101°42'E	Kuala Lumpar, Malaysia
3°08'S	60°01'W	Manaus, Brazil
8°58'N	79°31'W	Panama City, Panama
12°59'S	38°31'W	Salvador, Brazil

Special Instructions Objective: Ecosystem destruction
Students should find out about deforestation of rain forests. They should learn the reasons for it, how fast it is occurring, and how valuable the resource is. They should propose solutions that would satisfy all parties concerned.

Teacher's Guide
World Climatic Regions 7-3

Site: Taiga
Assignment Focus: Permafrost ecosystem
Students should find out about the delicate nature of this environment and should be able to report on how little it takes to set it back and hurt it beyond repair. Students should find examples of both wise and unwise interaction with this ecosystem.

64°34'N	40°32'E	Archangelsk, Russia
64°51'N	147°43'W	Fairbanks, Alaska
51°17'N	80°39'E	Moosonee, Canada
53°00'N	150°00'E	Okhotsk, Russia
65°00'N	27°00'E	Oulu, Finland
54°54'N	69°06'E	Petropavlovsk, Russia
47°34'N	52°43'W	St. John's, Canada
59°55'N	30°15'E	St. Petersburg, Russia
63°50'N	20°15'E	Ume, Sweden
62°27'N	114°21'W	Yellowknife, Canada

Special Instructions Objective: Way of life
Students should find out about family and community life in the Taiga region. They should find out about problems in the community associated with such a limited lifestyle. They should report on the general way of life in this region.

Teacher's Guide
World Climatic Regions 7-4

Site: Cold Temperate Forest
Assignment Focus: Resources, renewable
Students should find out about the renewable resources of the northern temperate forest. They should find out about businesses' efforts to increase the yield of the forest and how well those efforts are working. They should find out about government intervention and its effect.

44°49'N	68°47'W	Bangor, Maine
45°34'S	72°04'W	Coihaique, Chile
47°59'N	122°13'W	Everett, Washington
55°53'N	4°15'W	Glasgow, United Kingdom
54°17'N	31°00'E	Gorki, Russia
51°20'N	12°20'E	Leipzig, Germany
38°03'N	84°30'W	Lexington, Kentucky
43°03'N	141°21'E	Sapporo, Japan
37°50'N	112°37'E	Taiyuan, China
44°24'S	171°15'E	Timaru, New Zealand

Special Instructions Objective: Recreation, sites
Students should find out about recreational sites and determine criteria with which to categorize the sites. They should choose a site and design a facility that complements the environment and offers a variety of recreational opportunities.

Teacher's Guide
World Climatic Regions 7-5

Site: Warm Temperate Forest
Assignment Focus: Industrialization
Students should find out about this region and about the changes that take place through the seasons. They should assess the resources and recommend a facility designed to use the environment and the available resources most effectively.

36°52'S	174°45'E	Auckland, New Zealand
35°17'S	149°08'E	Canberra, Australia
23°07'N	113°18'E	Canton, China
25°25'S	49°15'W	Curitiba, Brazil
42°53'S	147°19'E	Hobart, Tasmania
28°03'N	81°57'W	Lakeland, Florida
32°50'N	83°38'W	Macon, Georgia
26°18'S	31°07'E	Mbabane, Swaziland
32°47'N	129°56'E	Nagasaki, Japan
31°25'S	26°52'E	Queenstown, South Africa

Special Instructions Objective: History
Students should find out about the patterns of population growth in these regions. They should find out why people want to live here and what other factors might cause such growth. They should make recommendations that would help prevent damage to the environment should the growth continue.

Teacher's Guide
World Climatic Regions 7-6

Site: Savanna
Assignment Focus: Local land use
Students should find out how the local population uses the savanna. Is it conservative use or not? Students should speculate about how much the local people depend on the land.

4°22'N	18°35'E	Bangui, Cen. Africa Republic
11°05'S	43°10'W	Barra, Brazil
15°56'S	50°08'W	Goiaz, Brazil
18°13'S	127°40'E	Halls Creek, W. Australia
17°50'S	31°10'E	Harare, Zimbabwe
0°19'N	32°35'E	Kampala, Uganda
20°44'S	139°30'E	Mount Isa, Queensland, Aus.
14°00'N	2°00'E	Niamey, Niger
27°50'S	64°15'W	Santiago del Estero, Arg.
22°13'N	97°51'W	Tampico, Mexico

Special Instructions Objective: Animals, food web
Students should find out about the animals of the savanna and report on the interdependence of the species that populate the region. They should make a picture of a typical food web for each region near each site.

Teacher's Guide
World Climatic Regions 7-7

Site: Mediterranean
Assignment Focus: Crops, new and old
Students should find out about the crops that are traditionally grown in the region. Which are the most important? They should also find out about any new crops that have come on the scene.

32°42'S	26°20'E	Adelaide, South Africa
36°47'N	3°30'E	Algiers, Algeria
41°38'N	41°38'E	Batumi, Georgia
33°53'N	35°30'E	Beirut, Lebanon
35°55'S	18°22'E	Cape Town, South Africa
37°55'N	22°53'E	Kórinthos, Greece
36°48'N	34°38'E	Mersin, Turkey
33°19'N	115°38'E	Bunbury, Western Australia
40°41'N	14°47'E	Salerno, Italy
34°03'N	118°15'W	Santa Barbara, California
37°23'N	5°59'W	Seville, Spain
33°02'S	71°38'W	Valparaiso, Chile

Special Instructions Objective: Way of life, business
Students should find out how people do business in this region—from the single entrepreneur to the large corporation. Students should look for patterns that could be interpreted as characteristic of the region.

Teacher's Guide
World Climatic Regions 7-8

Site: Desert Regions 1
Assignment Focus: Desert movement and growth
Students should find out how a desert moves, grows, and recedes, and about the factors that cause this flux. They should research ways that the population on the edge is attempting to stop the encroachment.

28°00'N	32°00'E	Arabian, Egypt
22°30'S	69°15'W	Atacama, Chile
28°30'N	106°00'W	Chihuahua, Mexico
36°30'N	117°00'W	Death Valley, California
24°30'S	126°00'E	Gibson, Australia
43°00'N	106°00'E	Gobi, Mongolia
33°00'N	57°00'E	Lut, Iran
23°00'S	15°00'E	Namib, Namibia
36°00'N	111°20'W	Painted Desert, Arizona
21°00'N	6°00'E	Sahara, Africa
32°00'N	40°00'E	Syrian, Middle East
27°00'N	70°00'E	Thar, India/Pakistan

Special Instructions Objective: Travel
Students should find out how best to travel on different desert terrains and should report on how to make travel a tolerable experience.

Teacher's Guide
World Climatic Regions 7-9

Site: Desert Regions 2
Assignment Focus: Resources, minerals
Students should find out about mineral resources other than oil, how available they are, how to get at them, and how much technology it would take to make mining them profitable.

21°30'S	125°00'E	Great Sandy, Australia
28°30'S	127°45'E	Great Victoria, Australia
23°00'S	22°00'E	Kalahari, southern Africa
39°00'N	60°00'E	Karakum, Turkmenistan
42°00'N	64°00'E	Kyzylkum, Kazakhstan
24°00'N	25°00'E	Libyan, Lybia
35°00'N	117°00'W	Mojave, southern California
20°30'N	33°00'E	Nubian, northeastern Sudan
18°00'N	45°00'E	Rub Al Khali, Arabia
37°59'N	120°23'W	Sonora, Az; Cal; Mexico
39°00'N	83°00'E	Taklimakan, China

Special Instructions Objective: Shelter
Students should find out the ways that the people in each site find or build shelter. They should look for similarities and the factors that drive the design decisions; then they should design a new type of shelter based on their new knowledge.

Teacher's Guide
World Climatic Regions 7-10

Site: Prairie
Assignment Focus: Grain production
Students should find out about the different types of grains grown in these regions. They should find out how hardy each is and how they are being developed to yield more harvest. Students should find out the nutritional information, growing seasons, ease of cultivation, and ease of harvest.

39°56'N	32°52'E	Ankara, Turkey
29°00'S	58°00'W	Corrientes, Argentina
48°27'N	34°59'E	Dnepropetrovsk, Ukraine
49°33'N	106°21'E	Darhan, Mongolia
32°45'N	97°20'W	Fort Worth, Texas
24°40'S	25°55'E	Gaborone, Botswana
30°59'S	150°15'E	Gunnedah, NSW, Australia
43°50'N	73°10'E	Karaganda, Kazakhstan
50°25'N	104°39'W	Regina, Canada
29°41'S	53°48'W	Santa Maria, Brazil
43°32'N	96°44'W	Sioux Falls, South Dakota

Special Instructions Objective: History
Students should find out about farming methods in these regions and about the farmers who use them. They should find out about traditional and new machinery, rotation, irrigation, and other facets.

8

SPECIFIC PHYSICAL GEOGRAPHY

Teacher's Guide

Teacher's Guide
Physical Geography 8-1

Site: Islands of the World
Assignment Focus: Site selection

Students should be surveying the many sites, looking at the political and environmental suitability of selected sites and the sites' proximity to strategic world locations.

50°N	3°W	Orkney Islands, N. Scotland
36°N	28°E	Rhodes, Aegean Sea, Turkey
4°N	73°30'E	Maldives, Indian Ocean
10°S	160°E	Guadalcanal, Solomon Is.,
22°N	158°W	Oahu, Hawaii, cen. Pacific
54°S	37°W	S. Georgia, S. Atlantic
13°N	59°30'W	Barbados, E. Caribbean
33°30'S	78°30'W	Robinson Crusoe Is., Chile
63°30'N	170°30'W	St. Lawrence Is., Bering Sea,
47°N	63°W	Prince Edward Is., Canada
29°N	119°W	Guadeloupe, N. Mexico,
32°45'N	17°W	Madeira Is., North Africa

Special Instructions Objective: Topographical maps

Students should produce detailed topographical drawings of the islands, carefully and neatly crafted, with keys and scales.

Teacher's Guide
Physical Geography 8-2

Site: Mountain Peaks
Assignment Focus: Accessibility, weather patterns

Students should find out the actual physical and political accessibility of the peaks. They should report on the prevailing weather patterns and also report on the equipment that would be necessary to explore the peaks.

32°39'S	70°00'W	Aconcagua, Argentina, Andes
45°50'N	6°52'E	Blanc, France, Alps
10°50'N	73°45'W	Cristóbal Colón, Colombia
43°21'N	42°26'E	Elbrus, Russia, Caucusus
27°59'N	86°56'E	Everest, Nepal, Himalayas
35°53'N	76°30'E	Pakistan, Himalayas
3°04'S	37°22'E	Kilimanjaro, Tanzania
0°10'S	37°20'E	Kirinyaga, Kenya
27°08'S	109°26'W	Kook, Chile
63°30'N	151°00'W	McKinley, Alaska
19°02'N	98°38'W	Popocatépetl, Mexico
31°03'N	7°55'W	Toubkal, Morocco, Atlas
78°35'S	85°25'W	Vinson Massif, Antarctica
1°45'S	133°25'E	Puncak Jaya, New Guinea

Special Instructions Objective: Local opinion

Students should find out how the local people feel about visitors on or near the peaks. They should be able to use the information to categorize the peaks by accessibility.

Teacher's Guide
Physical Geography 8-3

Site: Mountain Passes
Assignment Focus: Strategic geography

Students should find out about the local geography and the climate of the region around each pass. They should find out how the pass is used now and if it is significant or strategic and to whom.

20°23'N	18°18'E	Ahou, Tarso, Chad Tebesti
34°50'N	67°50'E	Bamian Pass, Hindu Kush
47°45'N	7°00'E	Belfort Gap
47°00'N	11°30'E	Brenner Pass, Austria to Italy
36°51'N	83°19'E	Cumberland Gap
39°19'N	120°20'W	Donner Pass, Sierra Nevada
45°50'N	8°10'E	Great St. Bernard
34°05'N	71°10'E	Khyber Pass
40°24'N	105°05'W	Loveland Pass, Rockies
38°31'N	73°41'E	Pereval, Akbajtal
43°01'N	1°19'W	Roncesvalles Sparto, France
42°22'N	108°55'W	South Pass, Rockies
46°30'N	8°30'E	St. Gotthard
38°48'N	22°32'E	Thermopylae, S. Greece

Special Instructions Objective: History, importance

Students should find out the history associated with each site and how the site was a factor in historical events. Students should report on the events that led to the use of the pass and the results. They should find out if the passes are still considered strategic locations and how the local populations use them.

Teacher's Guide
Physical Geography 8-4

Site: Atolls and Reefs
Assignment Focus: Environment, ecosystem

Students should find examples of reefs and atolls in the southwestern Pacific Ocean. They should find out about the ecosystems of the reefs and atolls that they choose and find evidence that could be used to argue against using the sites for weapons testing.

Special Instructions Objective: Shipwrecks/ Treasure hunts

Students should find out about the history of the dangers to ships near the sites and about efforts at salvage. They should find and report on actual accounts of shipwrecks and valuable finds.

Teacher's Guide
Physical Geography 8-5

Site: Glaciers
Assignment Focus: Resources, fresh water
Students should find out about the availability of fresh water. They should learn as much as they can, then propose ways to supply fresh water to needy nations.

Bossons, France	Bóver, Norway
Colombia, Alaska	Diablertz, Switzerland
Dinwoody, Wyoming	Fanarak, Norway
Gaupne, Norway	Grindelwald, Switzerland
Grinnel, Montana	Lambert, Antarctic
Lom, Norway	Mt. Rainier, WA, US
Nigards, Norway	Rhìne, Switzerland
Stechelberg, Switzerland	Taylor, Antarctic
Teton, Wyoming	Trient, Switzerland
Wright, Antarctic	Zermatt, Switzerland
Wilson Piedmont, Antarctic	

Special Instructions Objective: Dangers of exploration
Students should find out about explorations of the glacial regions and catalogue the dangers and the ways in which people have dealt with them. They should write a small handbook for explorers about any dangerous situation.

Teacher's Guide
Physical Geography 8-6

Site: Lakes
Assignment Focus: Products, fisheries
Students should find out about the geography of each site and investigate how fisheries are established and managed. They should determine if each site is a likely place for a fishery and recommend ways of establishing one.

45°00'N	60°00'E	Aral Sea, Russia
53°00'N	107°40'E	Baikal, Russia
42°00'N	50°30'E	Caspian Sea, Russia
42°08'N	80°04'W	Erie, United States
64°54'N	125°35'W	Great Bear, Canada
61°30'N	114°00'W	Great Slave, Canada
48°00'N	88°00'W	Lake Superior, US
12°00'S	34°30'E	Nyasa, Africa
6°00'S	29°30'E	Tanganyika, Africa
15°50'S	69°20'W	Titicaca, Bolivia
58°88'N	13°30'E	Vanârn, Sweden
1°00'S	33°00'E	Victoria, Africa

Special Instructions Objective: Way of life, economics
Students should find out how local populations live and how much their standards of living depend on the lake. Students should find out about the local economy and what impacts the lake has on the community and the family structure.

Teacher's Guide
Physical Geography 8-7

Site: Straits
Assignment Focus: Important shipping lanes
Students should find out about the importance of each site for shipping and how accessible or dangerous the site is. They should find out who controls each site and how friendly each government is to the shipping industry.

39°00'S	145°00'E	Bass Strait, Australia/Tasmania
65°30'N	169°00'W	Bering Strait, Alaska/Russia
41°00'N	29°00'E	Bosporus/Dardanelles, Turkey
34°40'N	129°00'E	Korea Strait, Korea/Japan
41°18'N	9°15'E	Strait of Bonifacio, Corsica/Sardinia
24°00'N	81°00'W	Strait of Florida, Florida/Cuba
35°57'N	5°36'W	Strait of Gibralter, Spain/Morocco
26°43'N	56°15'E	Strait of Hormuz, Iran/UAE
54°00'S	71°00'W	Strait of Magellan, Argentina
2°30'N	101°20'E	Strait of Malacca, Sumatra/Malaysia
24°00'N	119°00'E	Taiwan Strait, Taiwan/China
50°00'N	141°15'E	Tatar Strait, Sakhalin Island

Special Instructions Objective: Shelter
Students should find out about the life of a merchant marine. They should investigate the different jobs and who does them. They should find out what training is necessary for each job, what the requirements are for service, and how much each job pays.

Teacher's Guide
Physical Geography 8-8

Site: Capes 1
Assignment Focus: Lighthouses
Students should find out about lighthouses in the regions. They should report on the geography of each site and on the usefulness of each site for a locational facility, taking into account topography, population, proximity of high traffic shipping lanes, and accessibility. They should look into the history of lighthouses and the way that the LORAN system works.

20°46'N	17°03'W	Cape Blanc, Mauritania
8°38'N	104°44'E	Cape Ca Mau, VietNam
21°36'N	87°07'W	Cape Catoche, Mexico
55°59'S	67°16'W	Cape Horn, Chile
7°31'S	149°59'E	Cape Howe, Australia NSW
34°21'S	18°28'E	Cape of Good Hope, South Africa
4°22'N	7°44'W	Cape Palmas, Liberia
46°40'N	53°10'W	Cape Race, Newfoundland
37°55'N	16°04'E	Cape Spartivento, Italy
25°36'S	45°08'E	Cape St. Marie, Madagascar
58°37'N	5°01'W	Cape Wrath, United Kingdom
10°40'S	142°30'E	Cape York, Australia Queensland

Special Instructions Objective: Place names
Students should find out about the origins of the names of the capes—what they mean and why they were originally named. They should report on any anecdotes.

Teacher's Guide
Physical Geography 8-9

Site: Capes 2
Assignment Focus: Cargo ports, offshore

Students should find out about the geography of each site and then propose designs for offshore cargo facilities that would make use of the geographic characteristics of the area.

71°17'S	170°14'E	Cape Adare, Antarctica
35°00'S	136°00'E	Cape Catastrophe, S. Aus.
8°40'N	77°34'E	Cape Comorin, India
5°29'S	35°16'W	Cape de São Roque, Brazil
40°30'S	172°43'E	Cape Farewell, NZ
18°27'S	12°01'E	Cape Fria, Namibia
11°49'N	51°15'E	Cape Guardafuy, Somalia
35°13'N	75°32'W	Cape Hatteras, NC
50°52'N	156°40'E	Cape Lopatka, Russia,
40°25'N	124°25'W	Cape Mendocino, CA
21°57'S	43°16'E	Cape St. Vincent, Portugal
14°43'N	17°30'W	Cape Vert, Senegal

Special Instructions Objective: Names

Students should find out why the sites were given these names. They should propose new names for the sites based on present conditions and justify their choices.

Teacher's Guide
Physical Geography 8-10

Site: U.S. Capes
Assignment Focus: Common characteristics

Students should find out about the geography of each site and then look for common characteristics. They should synthesize information and profile a fictional site with typical physical and human characteristics. They should propose the design for an installation.

42°39'N	70°38'W	Cape Ann, Massachusetts
42°50'N	124°34'W	Cape Blanco, Oregon
28°30'N	80°35'W	Cape Canaveral, Florida
37°17'N	76°00'W	Cape Charles, Virginia
42°52'N	70°22'W	Cape Cod, Massachusetts
33°53'N	89°32'W	Cape Fear, North Carolina
48°32'N	124°43'W	Cape Flattery, Washington
35°13'N	75°32'W	Cape Hatteras, NC
38°56'N	74°54'W	Cape May, New Jersey
40°25'N	124°25'W	Cape Mendocino, CA
77°35'N	34°40'W	Cape Lookout, NC
25°12'N	81°05'W	Cape Sable, Florida
29°40'N	85°22'W	Cape San Blas, Florida

Special Instructions Objective: History, names

Students should find out why the capes were given these names. They should relate the names to historical events and write historical fiction that focuses on the naming of one of the sites that interests them.

Teacher's Guide
Physical Geography 8-11

Site: Active Volcanoes
Assignment Focus: Prediction

Students should find out about current methods used to predict eruptions and learn about new technology likely to be used. They should report the present state of the volcano, its last major eruption, and when it is expected to erupt again.

4°12'N	9°11'E	Cameroon, 1982, Cameroon
19°33'N	103°38'W	Colima, 1991, Mexico
37°50'N	14°55'E	Etna, 1990, Italy
9°59'N	83°51'W	Irazú, 1991, Costa Rica
12°10'S	44°15'E	Karthala, 1977, Comoro Is.
1°42'S	101°16'E	Kerinci, 1987, Sumatra
19°24'N	155°17'W	Kilauea, 1991, Hawaii
31°35'N	130°39'E	On-Take, 1991, Japan
60°29'N	152°45'W	Redoubt, 1990, Alaska

Others Include:	Sunguay, 1988, Equador
	Raung, 1990, Java
	Erebous, 1991, Antarctica
	Fuego, 1991, Guatemala
	Lascar, 1991, Chile

Special Instructions Objective: Local attitude

Students should find out whether people who live near volcanoes are fearful and how they handle those feelings. They should also find out what people do after an eruption has destroyed their homes.

Teacher's Guide
Physical Geography 8-12

Site: Historic Volcanoes
Assignment Focus: History, volcanic eruptions

Students should find out about the major eruption at each site and about the culture of the people that lived there when the volcano erupted. They should find out what effects the eruption had on the culture. Did people return to or abandon the site and why?

6°07'S	105°24'E	Krakatoa, Indonesia, 1883
37°50'N	14°55'E	Mt. Etna, Sicily, 1169, 1669
46°12'N	122°11'W	Mt. St. Helens, US, 1980
40°49'N	14°26'E	Mt. Vesuvius, 79, 1631
8°14'S	117°55'E	Tambora, Java, 1815

Others include:	Mt. Kelud, Java, 1966
	Mt. Lamington, NG, 1951
	Mt. Papanduyan, Java, 1772
	Mt. Pelée, Martinique, 1902
	Mt. Uzen-Dake, Japan, 1792
	Nevado del Ruiz, Col., 1985

Special Instructions Objective: Cultural tolerance

Students should find out about the cultures of the people who lived near the historic eruptions and find out what motivated them to stay and in some cases return to the site. They should speculate about attitudes that would allow a population to exist in the shadow of the threat of natural disaster.

Teacher's Guide
Physical Geography 8-13

Site: Rivers
Assignment Focus: Resources, minerals

Students should find out about the geography of these rivers, from the mouths to the sources. They should find out how the rivers are used and who uses them. They should also find out how clean and usable the rivers are.

0°10'S	49°00'W	Amazon, Brazil
52°56'N	141°10'E	Amur, China/Russia
31°48'N	121°10'E	Chang Jiang, China
6°04'S	12°24'E	Congo, Africa
45°30'N	29°45'E	Danube, Europe
72°25'N	126°40'E	Lena, Russia
10°15'N	105°55'E	Mekong, Southeast Asia
29°10'N	89°15'W	Mississippi, United States
38°50'N	90°08'W	Missouri, United States
31°20'N	31°00'E	Nile, Egypt
66°45'N	69°30'E	Ob, Russia
33°43'S	59°15'W	Parana, Argentina

Special Instructions Objective: Shelter

Students should find out about typical rural and urban life on the river and compare two examples. They should profile a typical relationship between people and a river environment.

Teacher's Guide
Physical Geography 8-14

Site: Sacred Mountains 1
Assignment Focus: Geographic commonalities

Students should find out about the mountains and the geographic settings of the mountains. They should look for common characteristics among the sites and report on those characteristics.

32°39'S	70°00'W	Aconcagua, Chile
39°40'N	44°24'E	Ararat, Egypt
54°10'N	449°40'E	Croagh Patrick, Ireland
3°04'S	37°22'E	Kilimanjaro, Tanzania
13°07'S	72°34'W	Machu Picchu, Peru
45°58'N	7°39'E	Matterhorn, Switzerland
32°25'S	116°18'E	Mt. Cooke, NZ
35°26'N	138°43'E	Mt. Fuji, Japan
45°55'N	68°55'W	Mt. Katahdin, Maine

Special Instructions Objective: Culture, folklore

Students should find out about stories about a sacred mountain at each site. They should relate two or three of those legends.

Teacher's Guide
Physical Geography 8-15

Site: Sacred Mountains 2
Assignment Focus: Unique characteristics

Students should find out about the geography of each mountain and look for distinguishing characteristics that would make the peak different from other peaks in the area or from the surrounding terrain. They should try to see the peak from the eyes of the people living nearby and report on what makes each mountain sacred.

63°30'N	151°00'W	Mt. McKinley, Alaska
40°05'N	22°21'E	Mt. Olympus, Greece
41°20'N	122°20'W	Mt. Shasta, California
28°32'N	33°59'E	Mt. Sinai, Egypt
31°46'N	35°14'E	Mt. Zion, Israel
38°51'N	105°03'W	Pikes Peak, Colorado
19°02'N	98°38'W	Popocatépetl, Mexico
7°58'S	113°35'E	Semeru, Java

Special Instructions Objective: Shelter

Students should find out about the cultures of the people who lived near the mountain when it was considered sacred. They should find out if the present local culture has the same belief system and, if not, what has changed their ideas.

Teacher's Guide
Physical Geography 8-16

Site: Coastal Islands
Assignment Focus: Government

Students should find out about the management of the islands. They should find out if the islands are independent or under the government of the nation whose coast they flank. They should look for differences in ways of life in the two types of administration and speculate why the differences exist.

4°30'N	9°30'E	Bioko Equa, Guinea
42°30'S	73°55'W	Chiloé, Chile
19°00'N	109°00'E	Hainan Dao, China
54°15'N	4°30'W	Isle of Man, UK
40°50'N	73°00'W	Long Island, New York
36°10'N	28°00'E	Rhodes
33°30'N	133°30'E	Shikoku, Japan
37°30'N	14°00'E	Sicily, Italy
49°16'N	123°07'W	Vancouver, BC
6°10'S	39°20'E	Zanzibar, Tanzania

Special Instructions Objective: Culture, mainland, island

Students should find out about the cultures of the island sites and compare them to the cultures of the mainland. They should look for significant differences or similarities and highlight and speculate on the reasons.

Teacher's Guide
Physical Geography 8-17
Site: Waterfalls
Assignment Focus: Accessibility
Students should find out about the geography of
each site and how navigable the approach to the site
is. They should learn the geological reasons for the
formation of the waterfall and note the similarities
in the geological data on these sites.

5°57'N	62°30'W	Angel, Venezuela
47°13'N	12°11'E	Krimml, Austria
44°48'S	167°44'E	Sutherland, New Zealand
55°30'N	126°00'W	Takkakaw, BC, Canada
29°14'S	31°30'E	Tugela, South Africa
35°28'N	119°33'W	Yosemite, California
	Others Include:	Cascade de Giétroz, Swit.
		Cuguenán, Venezuela
		East Mardalsfoss, Norway
		Gavarnie, France
		Great Falls, Guyana
		Ribbon Falls, California

Special Instructions Objective: History, discovery
Students should find out about the exploration and
discovery of each waterfall. They should find out
about the ways of life of the local populations and
the uses of the waterfalls. They should compare and
contrast those uses.

Teacher's Guide
Physical Geography 8-18
Site: Delta Regions
Assignment Focus: River and flood control
Students should find out how the people who live in
delta regions handle the river and work it for their
purposes. They should find out about attempts at flood
control and river rerouting. They should determine
if human efforts are environmentally sound or not.

0°10'S	49°00'W	Amazon, Brazil
15°50'N	95°06'E	Irrawaddy, Burma
45°30'N	29°45'E	Danube, Romania
23°20'N	90°30'E	Ganges, Bangladesh
69°00'N	136°30'W	Mackenzie, NWT, Canada
10°20'N	106°40'E	Mekong, VietNam
29°10'N	89°15'W	Mississippi, Louisiana
14°30'N	4°00'W	Niger interior, Mali
4°50'N	6°00'E	Niger mouth, Nigeria
31°20'N	31°00'E	Nile, Egypt
8°37'N	62°15'W	Orinoco, Venezuela
51°52'N	6°02'E	Rhine, Netherlands

**Special Instructions Objective: Culture and
animal life**
Students should find out about the adaptive facets
of the people who live in each of the regions. They
should also find out about the animal life that thrives
in these regions and how the population is affected
by the animal life.

Teacher's Guide
Physical Geography 8-19
Site: Ocean Trenches 1
Assignment Focus: Animal and plant life
Students should find out how much each of the sites
has been explored and what animal and plant life
thrives in the regions. Students should compare the
sites and report on similarities.

30°00'N	145°00'E	Bonin Trench
37°00'N	143°00'E	Japan Trench
30°00'S	177°00'W	Kermadez Trench
47°00'N	150°00'E	Kuril Trench
14°00'N	147°30'E	Mariana Trench
6°00'S	153°00'E	New Britain Trench
9°00'N	127°00'E	Philippine Trench
21°00'S	175°00'W	Tonga Trench
8°30'N	138°00'E	Yap Trench

Special Instructions Objective: Exploration
Students should find out about the history of
exploration at these sites. They should find out when
exploration started, who has led the way at different
times, and how the technology has improved to help
explorers be more thorough.

Teacher's Guide
Physical Geography 8-20
Site: Ocean Trenches 2
Assignment Focus: Effect on region
Students should find out about the regions that
surround an ocean trench and whether there are
any surface effects. They should find out about the
origin of the trenches and the tectonic mechanisms
that cause them.

51°00'N	179°00'E	Aleutian Trench
19°00'N	80°00'W	Cayman Trench
33°00'S	90°00'W	Chile Trench
10°30'S	110°00'E	Java Trench
15°00'N	95°00'W	Middle America Trench
20°00'S	168°00'E	New Hebrides Trench
6°30'N	134°30'E	Palau Trench
20°00'N	66°00'W	Puerto Rico Trench
25°45'N	128°00'E	Ryuku Trench
56°30'S	25°00'W	South Sandwich Trench

Special Instructions Objective: Waste disposal
Students should find out about the problem of waste
disposal. They should find out about present methods
and good and bad points of each method. Students
should find out if trenches are now being used and
should speculate about the pros and cons of using
them. They should be prepared to argue either side.

9

HUMAN-MADE GEOGRAPHY

Teacher's Guide

Teacher's Guide
Human-Made Geography 9-1

Site: Major Dams and Reservoirs
Assignment Focus: Statistics

Students should find out about the type, age, volume capacity, and construction of the dams on the list and the nature of the reservoir behind the dam. They should also learn about the administration of the dam and be able to discuss international control vs national control vs private control.

31°00'N	47°25'E	Atatürk, Firat, Turkey
56°14'N	117°17'W	Beunet WAC, Peace, Canada
15°43'S	32°42'E	Cabora Bassa, Zambezi,
51°30'N	68°19'W	Daniel Johnson, Manicouagan
47°45'N	106°50'W	Fort Peck, Montana, US
24°01'N	32°52'E	High Aswan, Egypt
36°00'N	114°27'W	Hoover, Colorado River, AZ,
	Others Include:	Grande Dixence, Dixence
		Mangla-Pakistan
		Rogun Nurek, Vakhsh, Tajikistan

Special Instructions Objective: Recreation

Students should find out about the recreational uses of the lakes that were created by the construction of the dams. They should also find out about the land acquisition at the time of the dam's construction. They should report any anecdotes about the process.

Teacher's Guide
Human-Made Geography 9-2

Site: Canals
Assignment Focus: History of construction

Students should find out about the history of the construction of the canals on the list. They should find out about the world political setting at the time, the reasons for the construction, the problems associated with the construction, and the past and present administration of the canals.

50°39'N	5°37'E	Albert, Belgium
51°57'N	5°25'E	Amsterdam, Rhine Netherlands
30°05'N	94°06'W	Beaumont, Port Arthur, TX
43°14'N	79°13'W	Welland, New York
54°20'N	10°08'E	Kiel, Germany
9°20'N	79°55'W	Panama, Panama
49°15'N	67°00'W	St. Lawrence Seaway, Québec
29°55'N	32°33'E	Suez, Egypt

Special Instructions Objective: Locks

Students should find out about locks and how and why they work. They should find out what training a canal worker needs, what the job was like in the past, and how it has changed as technology has improved.

Teacher's Guide
Human-Made Geography 9-3

Site: Rail Tunnels
Assignment Focus: Construction technology

Students should find out about the methods and the technology used to construct each of the tunnels. They should be able to see how technology improved as the demands for safer and more efficient methods increased.

44°29'N	11°20'E	Apennine, Bologna/Florence
47°08'N	10°12'E	Arlberg, Austria
55°20'N	3°27'W	Moffat, Colorado
45°15'N	6°54'E	Mont Cénis, French Alps
44°31'N	115°59'W	New Cascade, Washington
41°15'S	175°15'E	Rinautaka Wairarapa, NZ
41°40'N	140°55'E	Seikan, Tsugara Strait, Japan
46°15'N	8°00'E	Simplon, Alps, Switzerland/Italy
46°30'N	8°30'E	St. Gotthard, Swiss Alps
48°30'N	7°10'E	Vosges, France

Special Instructions Objective: Transportation, alternatives

Students should find out how goods and people were transported before the tunnel. They can speculate by looking at road maps or they can read accounts of the construction of the tunnels. They should figure how much travel time is saved by each tunnel.

Teacher's Guide
Human-Made Geography 9-4

Site: Vehicular Tunnels
Assignment Focus: Construction financing

Students should find out about the construction of the tunnels in general and then about the financing of the tunnel construction. If possible, they should find out who paid for them and who controls them. (In some cases, this information will be hard to find out.)

39°17'N	76°37'W	Fort McHenry, Baltimore
45°50'N	8°10'E	Great St. Bernard, Switzerland
37°02'N	76°23'W	Hampton Roads, Norfolk, VA
40°43'N	74°01'W	Lincoln, Hudson, New York/NJ
45°50'N	6°52'E	Mont Blanc, France
45°31'N	73°34'W	Mount Royal, Montreal
53°25'N	2°55'W	Queensbay, Mersey, Liverpool
46°30'N	8°30'E	St. Gotthard, Switzerland

Special Instructions Objective: History

Students should find out about the new opportunities that opened up with the use of the new tunnel. They should look for possibilities in the commercial, recreational, and personal realms.

Teacher's Guide
Human-Made Geography 9-5
Site: Ancient Ruins
Assignment Focus: Archaeological findings
Students should find out about archaeological
excavations. They should find out what we have
learned and who is in charge of these excavations.

13°26'N	103°52'E	Angkor Cambodia
32°32'N	44°25'E	Babylon, Iraq
20°40'N	88°35'W	Chichén Itzá, Yucatán
35°18'N	25°10'W	Knosós, Crete
15°30'N	45°21'E	Ma'rib, Yemen (Sheba)
13°07'S	72°34'W	Machu Picchu, Peru
6°55'N	158°15'E	Nan Matol, Ponapé Caroline Is.
30°03'N	31°15'E	Pyramids, Cairo
27°07'S	109°22'W	Rapa Nui, Easter Island
51°11'N	1°49'W	Stonehenge, Salisbury Plain
17°20'N	89°39'W	Tikal, Guatemala
50°41'N	4°46'W	Tintagel, Cornwall, UK (Arthur)
39°57'N	26°15'E	Troy, Turkey
20°16'S	30°55'E	Zimbabwe Ruins (Solomon)

Special Instructions Objective: Culture, fiction
Students should find out about the people who lived
near the site when it was a cultural center. They
should write a fictional account of a visit to the site,
including information about all facets of the culture.

Teacher's Guide
Human-Made Geography 9-6
Site: Airports
Assignment Focus: Accessibility
Students should find out about the layout of the
facilities and the reasons each was built in that way.
They should speculate why the site has become a hub
for air traffic and cite the good and bad points of each
site. They should propose a new site.

33°45'N	82°43'W	Atlanta, United States
48°52'N	2°20'E	Charles DeGaulle, Paris
32°47'N	96°48'W	Dallas/Ft. Worth, US
42°45'N	12°29'E	Fiumicino, Rome
52°21'N	14°33'E	Frankfurt, Germany
51°30'N	0°10'W	Gatwick, London
35°40'N	139°46'E	Haneda, Tokyo
51°30'N	0°10'W	Heathrow, London
22°15'N	114°10'E	Hong Kong
34°03'N	118°15'W	LAX, Los Angeles
43°39'N	79°23'W	Lester, Pearson, Toronto
35°40'N	139°46'E	Narita, Tokyo
41°53'N	87°38'W	O'Hare, Chicago
48°52'N	2°20'E	Orly, Paris
35°37'N	137°14'E	Osaka, Japan

Special Instructions Objective: History
Students should find out about the job of air traffic
controller. They should learn about training and daily
requirements of the job. They should find out what
makes the job difficult and what makes it easier.

Teacher's Guide
Human-Made Geography 9-7
Site: Bridges
Assignment Focus: Construction types
Students should find out about the different types of
bridges represented. They should learn how and why
each type was built and how well the bridges fit in
with the environment.

49°16'N	123°07'W	Alex Fraser, Vancouver, BC
41°09'N	8°37'W	Arrábida, Porto, Portugal
46°11'N	123°50'W	Astoria, Oregon
41°00'N	29°0'E	Bosporus, Istanbul
13°32'N	100°36'E	Chao Phraya, Thailand
37°49'N	122°29'W	Golden Gate, San Francisco, CA
22°35'N	88°20'E	Howrah, Calcutta
53°40'N	0°10'W	Humber, Hull UK
33°56'N	130°57'E	Kanmon Strai, Kyushu-Honshu,
45°48'N	16°00'E	KRK, Zagreb, Croatia
32°48'N	79°57'W	Mark Clark Expressway, SC
35°37'N	137°14'E	Minato, Osaka, Japan
46°49'N	71°13'W	Pierre LaPort, Québec, Canada
68°26'N	17°25'E	Skjomen, Narvik, Norway
33°52'S	151°13'E	Sydney Harbor, Australia
9°20'N	79°55'W	Thatcher Ferry, Panama Canal
58°00'N	11°38'E	Tjïrn, Sweden
40°43'N	74°01'W	Verrazano Narrows, NY

Special Instructions Objective: History, construction
Students should find out about the building of the
bridges, the various jobs, and the architects who
designed the spans.

Teacher's Guide
Human-Made Geography 9-8
Site: Wonders of the World
Assignment Focus: Site characteristics
Students should find out about the nature of the sites.
They should be able to determine the characteristics
that would allow each site to be included in the list.
They should find out how long the site lasted and
what evidence is left of the site.

36°10'N	28°00'E	Colossus of Rhodes
32°32'N	44°25'E	Hanging Gardens of Babylon
37°02'N	27°06'E	Mausoleum at Halicarnassus
31°12'N	29°54'E	Pharos of Alexandria
40°05'N	22°21'E	Statue of Zeus at Olympia
37°55'N	27°20'E	Temple of Artemis, Ephesus
30°03'N	31°15'E	The Pyramids of Egypt, Cairo

Special Instructions Objective: History, construction
Students should find out how these were constructed
and who designed them, who was in charge of building them, and who actually did the construction.
They should find out what the site meant to the local
culture.

10
HISTORICAL
PERSPECTIVE

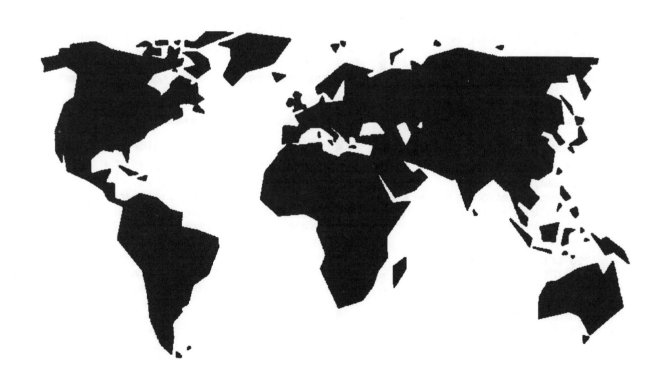

Teacher's Guide

Teacher's Guide
Historical Perspective 10-1

Site: United States, 1870 to 1900
Assignment Focus: Post–Civil War period

Students should be looking for and reporting on the environmental, geographic, economic, and strategic factors that caused these cities to be established and to thrive during the period of westward expansion.

39°06'N	84°31'W	Cincinnati, Ohio
39°05'N	94°35'W	Kansas City, Missouri
35°08'N	90°03'W	Memphis, Tennessee
45°33'N	122°36'W	Portland, Oregon
38°35'N	121°30'W	Sacramento, California
38°38'N	90°11'W	Saint Louis, Missouri
40°46'N	111°53'W	Salt Lake City, Utah
29°28'N	98°31'W	San Antonio, Texas
35°41'N	105°56'W	Santa Fe, New Mexico
37°41'N	97°20'W	Wichita, Kansas

Special Instructions Objective: Period trail map

Students should research old maps, their uses, and how they satisfied the needs of the people who used them. Students should produce a detailed trail map as it would have appeared in that time period, with notes on hazards, shortcuts, seasonal information, river fords, and so on.

Teacher's Guide
Historical Perspective 10-2

Site: Early Farming Cultures
Assignment Focus: Farming methods

Students should find out about farming methods and crops that were characteristic of each site. They should also find out what the attraction of each site was.

20°N	100°W	Central Mexico
30°N	30°E	Egypt
39°N	22°E	Greece
45°N	35°E	Iraq/Iran
40°N	110°E	Northeast China
30°N	70°E	Pakistan
5°S	75°W	Peru
15°N	100°E	Thailand
38°N	35°W	Turkey
10°N	0°	West Africa

Special Instructions Objective: Development of civilization

Students should find out about the change from hunting-gathering at each site and about the early development of civilization. They should find out about the settlement structure, from the family unit to the overall management or governmental unit.

Teacher's Guide
Historical Perspective 10-3

Site: The Roman Empire, CE 120
Assignment Focus: Evidence of Roman influence

Students should find evidence of Roman culture in the named sites. They should look for evidence in architecture, people, and language.

31°12'N	20°54'E	Alexandria
36°14'N	36°07'E	Antioch
37°59'N	23°44'E	Athens
47°30'N	19°05'E	Budapest
36°49'N	10°18'E	Carthage
32°48'N	21°59'E	Cyrene
31°46'N	35°14'E	Jerusalem
38°43'N	9°08'W	Lisbon
51°30'N	0°10'W	London
45°45'N	4°51'E	Lyon
43°18'N	5°24'E	Marseilles
48°52'N	2°20'E	Paris
42°45'N	12°29'E	Rome
37°04'N	15°18'E	Syracuse
39°50'N	4°00'W	Toldeo
48°12'N	16°22'E	Vienna

Special Instructions Objective: Roman characteristics

Students should find out about the unique nature of being a Roman citizen in each of the sites and what general characteristics of being Roman were common to all of the sites.

Teacher's Guide
Historical Perspective 10-4

Site: Middle Ages Trade Routes
Assignment Focus: Goods and services

Students should find out about the goods and services that were part of the trading culture of the early Middle Ages. They should find out what was manufactured and where, and about the means of exchange.

31°12'N	20°54'E	Alexandria
51°13'N	3°14'E	Bruges
41°01'N	28°58'E	Constantinople
44°25'N	8°57'E	Genoa
53°33'N	10°00'E	Hamburg
50°26'N	30°31'E	Kiev
51°30'N	0°10'W	London
43°18'N	5°24'E	Marseilles
40°50'N	14°15'E	Naples
58°31'N	31°17'E	Novgorod
48°52'N	2°20'E	Paris
43°43'N	10°23'E	Pisa
42°45'N	12°29'E	Rome
59°20'N	18°03'E	Stockholm
45°27'N	12°21'E	Venice

Special Instructions Objective: Trade transportation

Students should find out about modes of transportation that were in use. They should find out about overland transportation and marine transportation.

Teacher's Guide
Historical Perspective 10-5
Site: The Crusades, 1100–1200
Assignment Focus: Objectives of the Crusades
Students should find out about the Crusades. They should also find out how people traveled and if they accomplished their goals. Students should report on the part that each site played.

36°14'N	36°07'E	Antioch
41°01'N	28°58'E	Constantinople
44°25'N	8°57'E	Genoa
31°46'N	35°14'E	Jerusalem
45°45'N	4°51'E	Lyon
43°18'N	5°24'E	Marseilles
49°08'N	6°10'E	Metz
48°52'N	2°20'E	Paris
43°43'N	10°23'E	Pisa
42°45'N	12°29'E	Rome
34°26'N	35°51'E	Tripoli
45°27'N	12°21'E	Venice
48°12'N	16°22'E	Vienna
39°55'N	37°48'E	Zara

Special Instructions Objective: The Crusaders
Students should find out about the people that went on the Crusades and what motivated them. Students should find out about the leaders, what they were thinking, and what their objectives were. They should find out what it was like to live during that era.

Teacher's Guide
Historical Perspective 10-6
Site: Barbarian Europe, Middle Ages
Assignment Focus: Tribal territory
Students should find out about the following tribes of Barbarians that swept down on Europe and helped bring about the fall of the Roman Empire. They should find out who they were, what territory they controlled, when they controlled it, and what their ambitions were.

Anglo-Saxons	Basques
Burgundians	Byzantines
Franks	Huns
Jutes	Magyars
Muslims	Ostrogoths
Picts	Scots
Slavs	Vandals
Vikings	Visigoths

Special Instructions Objective: Way of life, leadership
Students should find out about the way of life of each of the tribes, who their leaders were, and what they wanted for their people. Personal sketches of the leaders would enhance students' study of the people.

Teacher's Guide
Historical Perspective 10-7
Site: British Empire 1, 1850–1900
Assignment Focus: Colonial resources
Students should investigate the resources of the colonies and find out what strategic importance each played in the empire.

25°00'S	135°00'E	Australia
17°35'N	88°35'W	Belize
22°00'S	24°00'E	Botswana
60°00'N	95°00'W	Canada
12°00'N	15°00'W	Guinea-Bissa
5°00'N	59°00'W	Guyana
13°30'S	34°00'E	Malawi
41°00'S	174°00'E	New Zealand
10°00'N	8°00'E	Nigeria
30°00'S	26°00'E	South Africa
8°30'N	11°30'W	Sierra Leone
8°00'N	1°10'E	Togo
15°00'S	30°00'E	Zambia
20°00'S	30°00'E	Zimbabwe

Special Instructions Objective: Colonial government
Students should find out how the British kept peace in the colonies. They should find out how the colonies were governed and what was needed to keep them productive. Students should also find out about the causes of the empire's deterioration.

Teacher's Guide
Historical Perspective 10-8
Site: British Empire 2, 1850–1900
Assignment Focus: Purpose of imperialism
Students should find out about the acquisition of the sites and their inclusion in the British Empire. They should find out who was in power, what the govern-mental policy was, and how imperialism was justified.

22°00'N	98°00'E	Burma
27°00'N	30°00'E	Egypt
22°00'N	77°00'E	India
1°00'N	38°00'E	Kenya
4°00'N	102°00'E	Malaysia
21°00'N	57°00'E	Oman
30°00'N	70°00'E	Pakistan
6°00'S	150°00'E	Papua New Guinea
10°00'N	49°00'E	Somalia
15°00'N	30°00'E	Sudan
43°00'S	147°00'E	Tasmania
1°02'N	32°00'E	Uganda
15°00'N	44°00'E	Yemen

Special Instructions Objective: British influence
Students should find out about continued British influence in the former colonies. They should determine if it is stronger in some places and weaker in others, and they should speculate as to why the discrepancies existed.

Teacher's Guide
Historical Perspective 10-9

Site: Exploration of the New World, 1490–1750
Assignment Focus: Explorers and their sponsors
Students should find out what explorers (not native to the sites) explored these places. They should find out who these explorers represented and what their sponsors' objectives were.

6°15'S	78°50'W	Cajamarca
46°00'N	60°30'E	Cape Breton Island
13°31'S	71°59'W	Cuzco
42°20'N	83°03'W	Detroit
45°00'N	87°30'W	Green Bay
45°31'N	73°34'W	Montreal
29°58'N	90°07'W	New Orleans
9°00'N	80°00'W	Panama
46°49'N	71°13'W	Québec
32°43'N	117°09'W	San Diego
13°42'N	89°12'W	San Salvador
29°51'N	81°25'W	St. Augustine

Special Instructions Objective: Native cultures
Students should find out what the explorers' goals were. They should find out about the cultures the explorers encountered and how the explorers interacted in those cultures. Students should be able to suggest why some got along with the native peoples and some didn't.

Teacher's Guide
Historical Perspective 10-10

Site: World War I, Europe, 1914–1918
Assignment Focus: Site importance
Students should find out about each of the sites and what role it played in the progression and outcome of the war. They should find out about the conflicts that led to the war and how the war changed the face of Europe.

49°54'N	2°18'E	Amiens
54°05'N	8°50'E	Cantigny
49°03'N	3°24'E	Chateau Thierry
40°03'N	17°58'E	Gallipoli
53°57'N	9°00'E	Marne
53°45'N	21°45'E	Masurian Lakes
48°54'N	5°33'E	Saint-Mihiel
43°50'N	18°25'E	Sarajevo, Bosnia
49°55'N	2°30'E	Somme
59°55'N	30°15'E	St. Petersburg
49°10'N	5°23'E	Verdun

Special Instructions Objective: Innovations in warfare
Students should find out about innovations in warfare techniques and weaponry. They should find out who invented the new methods and how they worked.

Teacher's Guide
Historical Perspective 10-11

Site: World War II, Pacific Theater
Assignment Focus: Battle sites
Students should find out about the sites listed and about their roles in the progress of the war. They should find out who was in charge at each site, how long the fighting lasted, and what the cost was.

20°30'N	121°50'E	Bataan
20°00'S	158°00'E	Coral Sea
14°35'N	121°00'E	Corregidor
9°32'S	160°12'E	Guadalcanal
13°28'N	144°47'E	Guam
34°24'N	132°27'E	Hiroshima
24°47'N	141°20'E	Iwo Jima
10°50'N	124°50'E	Leyte Gulf
28°13'N	177°22'W	Midway
32°47'N	129°56'E	Nagasaki
26°20'N	127°47'E	Okinawa
21°20'N	158°00'W	Pearl Harbor

Special Instructions Objective: Culture of war
Students should find out about the American reaction to the war. They should look for examples of glamorization and the downplaying of its full effect. They should also look for attempts to report it realistically.

Teacher's Guide
Historical Perspective 10-12

Site: World War II, European Theater
Assignment Focus: Industry
Students should find out about the role each site played in the Allied or Axis plans and in the war. They should write a chronology of the war that includes each site.

48°54'N	5°33'E	Algiers
43°50'N	18°25'E	Anzio
49°55'N	2°30'E	Bastogne
59°55'N	30°15'E	Casablanca
49°10'N	5°23'E	Cologne
42°29'N	79°21'W	Dunkirk
30°99'N	28°57'E	El Alamein
59°55'N	30°15'E	Leningrad
51°30'N	0°10'W	London
36°00'N	0°35'W	Oran
48°52'N	2°20'E	Paris
50°34'N	7°14'E	Remagen
42°45'N	12°29'E	Rome
48°44'N	44°25'E	Stalingrad
32°05'N	23°59'E	Tobruk
34°00'N	9°00'E	Tunis

Special Instructions Objective: History
Students should find out about the plans of the Axis and the Allies and create a time line showing the interplay. They should find out about the mistakes and good fortune.

11
WORLD
EXPLORATION

Teacher's Guide

Teacher's Guide

World Exploration 11-1

Explorer: Henry Hudson

Henry Hudson explored during 1607–1611. He searched for a northeast and then a northwest passage. He is known for exploring regions of the northeast United States and Canada and the Hudson Bay, where he fell victim to a mutiny. Sites are student researched and chosen.

Teacher's Guide

World Exploration 11-2

Explorer: Ibn Battuta

Ibn Battuta explored northern and eastern Africa, the Middle East, central Asia, India, and southeastern Asia during 1340–1360. He dictated a detailed account that was the standard for knowledge of the Orient for centuries.

39°00'N	35°00'E	Anatolia
23°07'N	113°18'E	Canton
42°30'N	45°00'E	Caucasus
45°00'N	34°00'E	Crimea
22°00'N	77°00'E	India
21°29'N	39°12'E	Jiddah
9°17'S	28°20'E	Kilwa
21°27'N	39°49'E	Mecca
4°03'S	39°40'E	Mombasa

Teacher's Guide

World Exploration 11-3

Explorers: Richard Burton and John Speke

Burton and Speke explored western equatorial Africa during the late 1850s. They came upon Lake Victoria and later disagreed about the discovery of the sources of the White and Blue Niles.

2°30'S	32°54'E	Mwanza
1°01'N	32°57'E	Namasagali
5°20'S	32°30'E	Tabora
4°55'S	29°41'E	Ujiji
6°10'S	39°20'E	Zanzibar

Teacher's Guide

World Exploration 11-4

Explorers: Henry Stanley and David Livingstone

David Livingstone was a missionary in South Africa. From there he explored central and western Africa during the 1850s and 1860s. He reached Victoria Falls and many of the interior rivers. His explorations inspired many other expeditions to Africa.

5°51'S	13°03'E	Boma
12°00'S	34°30'E	Lake Malawi
20°37'S	22°40'E	Lake Ngami
12°19'S	30°18'E	Livingstone Memorial
0°26'N	25°20'E	Lualaba River
5°20'S	32°30'E	Tabora
4°55'S	29°41'E	Ujiji
17°56'S	25°50'E	Victoria Falls
6°04'S	12°24'E	Zaire River
6°10'S	39°20'E	Zanzibar

Teacher's Guide

World Exploration 11-5

Explorer: Hsuan-tsang

Hsuan-tsang made a sixteen-year pilgrimage from China to India, beginning in 629. His accounts were well received in China upon his return and he became a court translator.

25°27'N	81°51'E	Allahabad
25°00'N	93°00'E	Assam
39°55'N	116°23'W	Beijing
43°00'N	106°00'E	Gobi Desert
35°00'N	71°00'E	Hindu Kush
24°20'N	67°47'E	Indus River
34°00'N	76°00'E	Kashmir

Teacher's Guide

World Exploration 11-6

Explorer: Alexine Tinne

Alexine Tinne was a Dutch heiress who explored the Nile River in the early 1860s with her mother, Harriet Tinne. She traveled only in the highest style and luxury and only for the pleasure of going where no one had preceded her.

27°10'N	31°16'E	Asyut
30°03'N	31°15'E	Cairo
24°05'N	32°53'E	Aswan
15°50'N	33°00'E	Khartoum
52°06'N	4°18'E	The Hague

Teacher's Guide

World Exploration 11-7

Explorer: Mary Kingsley

Mary Kingsley explored the coast of West Africa in the 1890s. She went to study the zoology of the region and the religious beliefs of the native peoples. She became an advocate of change in Britain's imperial policy in Africa.

28°00'N	15°30'W	Canary Island
4°12'N	9°11'E	Mt. Cameroun
0°49'S	9°00'E	Ogooué River

Teacher's Guide

World Exploration 11-8

Explorer: May French Sheldon

May French Sheldon was an American explorer who explored eastern Africa in the 1890s. She visited the dreaded Masai and won their friendship. She proved that if one came to the native people in friendship and understanding, the people generally accept the expedition.

7°42'S	30°00'E	Marunga
4°03'S	39°40'E	Mombasa
3°04'S	37°22'E	Mt. Kilimanjaro
5°26'S	38°58'E	Pangani
3°24'S	37°41'E	Taveta
6°10'S	39°20'E	Zanzibar

Teacher's Guide

World Exploration 11-9

Explorer: Delia Denning Akeley

Delia Denning Akeley was an American explorer and the first Western woman to cross the African continent. Her explorations from 1905 to 1929 were to collect biological samples and to learn about the people. She was the first outsider to make contact with the Pygmy people and she brought back many artifacts as well as a great understanding of the continent of Africa.

43°28'N	88°50'W	Beaver Dam, Wisconsin
3°14'S	36°45'E	Mount Meru, Kenya
		Ituri Forest, Zaire
		Muddo Gashi
		Pygmys
		San Kuri
		Tama River, Kenya

Teacher's Guide

World Exploration 11-10

Explorer: Dame Freya Madeline Stark

Dame Stark did not explore for honor, wealth, or fame but because she had the heart of an explorer. Her explorations centered in the Middle East from the 1930s to the 1950s. She wrote extensively of her travels and was given many honors as an explorer.

12°46'N	45°01'E	Aden
33°21'N	44°23'E	Baghdad
30°03'N	31°15'E	Cairo
44°28'N	7°22'E	Dronero Italy
44°39'N	63°36'W	Halifax, Canada
33°30'N	48°40'E	Lorestan
36°00'N	54°00'E	Mazandaran
48°52'N	2°20'E	Paris

Teacher's Guide

World Exploration 11-11

Explorer: Gertrude Bell

Gertrude Bell explored the Middle East from 1900 to 1914. She learned to speak, write, and read Persian and Arabic and became immersed in the culture of the region. She worked in the Middle East all her life and foresaw many of the changes that have taken place there.

36°12'N	37°10'E	Aleppo
33°21'N	44°23'E	Baghdad
54°45'N	1°40'W	County Durham, England
33°30'N	36°15'E	Damascus
46°32'N	8°08'E	Finsteraarhorn Alps
27°33'N	41°42'E	Ha'il, Arabia
33°38'N	42°49'E	Hit
31°46'N	35°14'E	Jerusalem
32°30'N	43°45'E	Karbala
37°52'N	32°31'E	Konya
34°33'N	38°17'E	Palmyra, Turkey
30°19'N	35°29'E	Petra, Turkey
35°00'N	38°00'E	Syria
35°40'N	51°26'E	Tehran, Iran

Teacher's Guide

World Exploration 11-12

Explorer: Alexandra David-Neel

Alexandra David-Neel was a French journalist who, at the age of 51, was the first Western woman to penetrate the forbidden city of Lhasa, Tibet. She explored the mountains of China and Tibet in 1923 and 1924 and wrote of her travels in books that are thrilling even by today's standards.

22°00'N	98°00'E	Burma
35°00'N	105°00'E	China
14°40'N	17°26'W	Dakar Pass
44°06'N	6°14'E	Digne France
38°00'N	137°00'E	Japan
29°42'N	91°07'E	Lhasa
27°50'N	88°30'E	Sikkim
44°52'N	12°30'E	Source of Po

Teacher's Guide

World Exploration 11-13

Explorer: Marco Polo

Marco Polo explored China and India from 1270 through 1290. He traveled first with his family and then again by himself. He became a favorite of the ruler of China, and only by going on a diplomatic mission was he able to return to the West.

40°01'N	32°21'E	Ayas
33°18'N	44°23'E	Baghdad
40°23'N	49°51'E	Baku
39°55'N	116°23'W	Beijing
41°01'N	28°58'E	Constantinople
27°58'N	116°20'E	Fuzhou
44°25'N	8°57'E	Genoa
26°34'N	56°15'E	Hormuz
31°46'N	35°14'E	Jerusalem
39°29'N	75°58'E	Kashi
4°00'N	102°00'E	Malaya
7°40'N	80°50'E	Sri Lanka
0°01'N	102°00'E	Sumatra
45°27'N	12°21'E	Venice

Teacher's Guide

World Exploration 11-14

Explorers: Willem Barents, Vitus Bering, Fridtjof Nansen, and Robert Peary

Willem Barents (1594), Vitus Bering (1740s), Fridtjof Nansen (1890s), and Robert Peary (1890s) were all explorers of the frozen Arctic Ocean region and the North Pole. Each contributed a great deal to the understanding of the region.

74°30'N	19°00'E	Bear Island
65°30'N	169°00'W	Bering Strait
78°19'N	72°38'W	Etah
81°00'N	55°00'E	Franz Josef Land
90°00'N	0°00'	North Pole
74°00'N	57°00'E	Novaya Zemlya

Teacher's Guide

World Exploration 11-15

Explorer: Christopher Columbus

Christopher Columbus set out to find a western route to the East and its spices and riches. He instead ended up exploring the Caribbean and its islands during his four journeys in the 1490s.

21°30'N	80°00'W	Cuba
19°00'N	71°00'W	Hispaniola
18°15'N	77°30'W	Jamaica
18°15'N	66°30'W	Puerto Rico
13°42'N	89°12'W	San Salvador

Teacher's Guide

World Exploration 11-16

Explorers: Meriwether Lewis and William Clark

Meriwether Lewis and William Clark were commissioned by President Thomas Jefferson to explore the Louisiana Territory in 1804. Their explorations are well documented and a prime example of the rigors of exploration.

Teacher's Guide

World Exploration 11-17

Explorer: James Cook

Captain James Cook explored the South Pacific, Australia, and New Zealand. His innovative knowledge of long journeys and sailing were instrumental in the success of his explorations.

10°30'S	105°40'E	Christmas Island
41°20'S	174°25'E	Cook Strait
27°07'S	109°22'W	Easter Island
19°10'S	149°00'E	Great Barrier Reef
24°00'N	167°00'W	Sandwich Island (Hawaii)
17°37'S	149°27'W	Tahiti
20°00'S	175°00'W	Tonga

Teacher's Guide

World Exploration 11-18

Explorers: Robert Burke, William Wills, and John Stuart

Robert Burke and William Wills, as well as John Stuart, explored Australia in the 1860s. The two explorations went from the south to the north and back again. Burke and Wills died before reaching their goal. Stuart's trip was successful.

32°42'S	26°20'E	Adelaide
28°29'S	137°46'E	Cooper Creek
12°28'S	130°50'E	Darwin
17°44'S	139°22'E	Gulf of Carpathia
37°49'S	144°58'E	Melbourne
32°24'S	142°26'E	Menindee
25°00'S	137°00'E	Simpson Desert

Teacher's Guide

World Exploration 11-19

Explorers: Ferdinand Magellan and Sebastian del Cano

Ferdinand Magellan set out to circumnavigate the globe by sailing west from Spain in 1519 with 260 men and five ships. Magellan was killed before he reached his goal, but a captain, Sebastian del Cano, made it home with one ship and few survivors in 1522.

28°00'N	15°30'W	Canary Islands
34°21'S	18°28'E	Cape of Good Hope
9°00'N	168°00'E	Marshall Islands
2°00'S	128°00'E	Moluccas
44°00'S	68°00'W	Patagonia
13°00'N	122°00'E	Philippines
40°00'N	4°00'W	Spain
54°00'S	69°00'W	Tierra del Fuego

Teacher's Guide

World Exploration 11-20

Explorer: First People on the Mountain

Each tall peak on each of the continents has a story of its conquest, beginning with Emperor Hadrian in the first century and continuing to the 1960s. Both men and women have been involved in the conquests on all of the continents.

32°39'S	70°00W	Aconcagua, Argentina, Andes
45°50'N	6°52'E	Blanc, France, Alps
10°50'N	73°45'W	Cristóbal Colón, Colombia
27°59'N	86°56'E	Everest, Nepal, Himalayas
35°53N	76°30'E	K2 Pakistan, Himalayas
3°04'S	37°22'E	Kilimanjaro, Tanzania
38°51'N	105°03'W	Pike's Peak, Colorado, US
27°08'S	109°26'W	Cook, New Zealand
63°30'N	151°00'W	McKinley, Alaska
19°02'N	98°38'W	Popocatépetl, Mexico
78°35'S	85°25'W	Vinson Massif, Antarctica
45°58'N	7°39'E	Matterhorn, Switzerland

MAPS

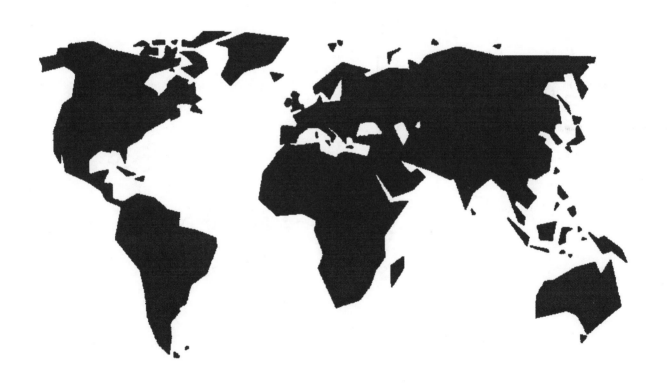

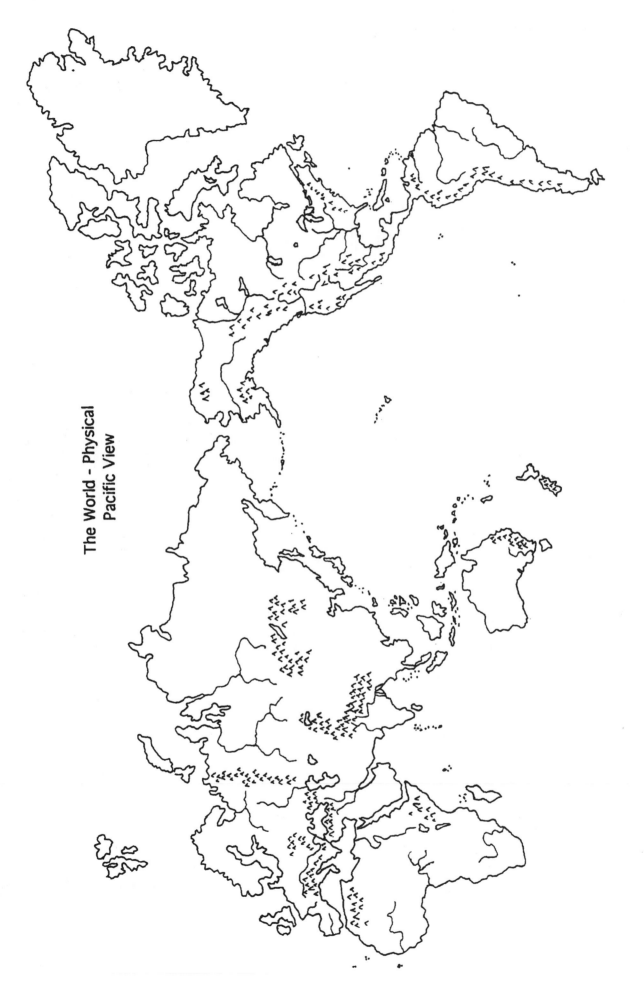

The World – Physical
Pacific View

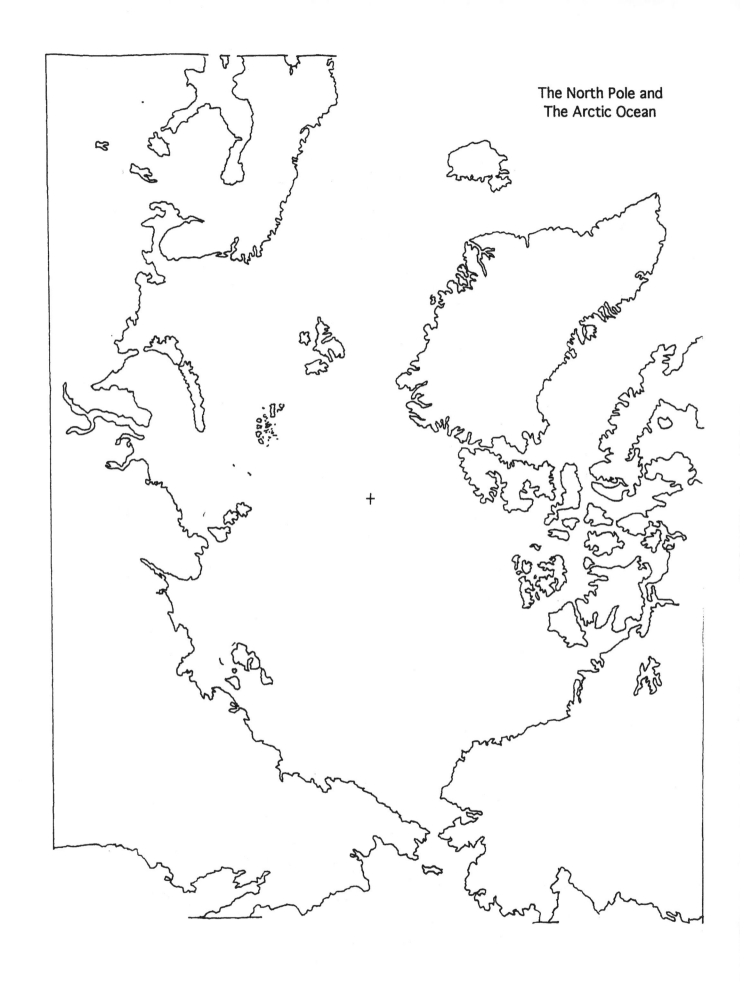

Antarctica and The South Pole

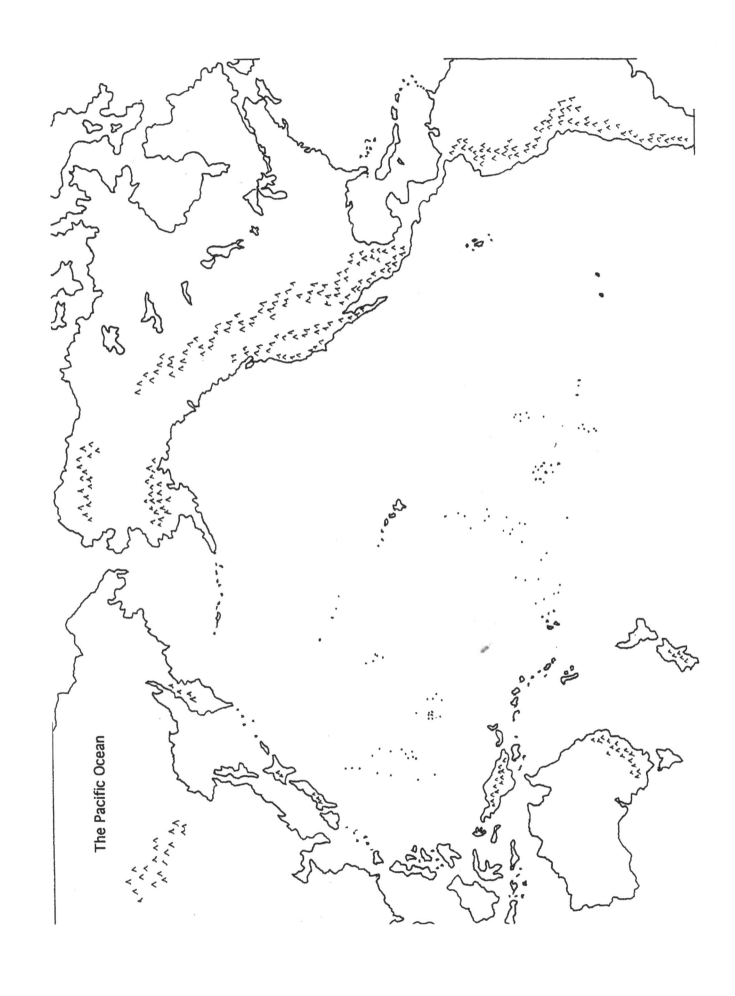

The Pacific Ocean

The Atlantic Ocean

The Western Hemisphere

North America

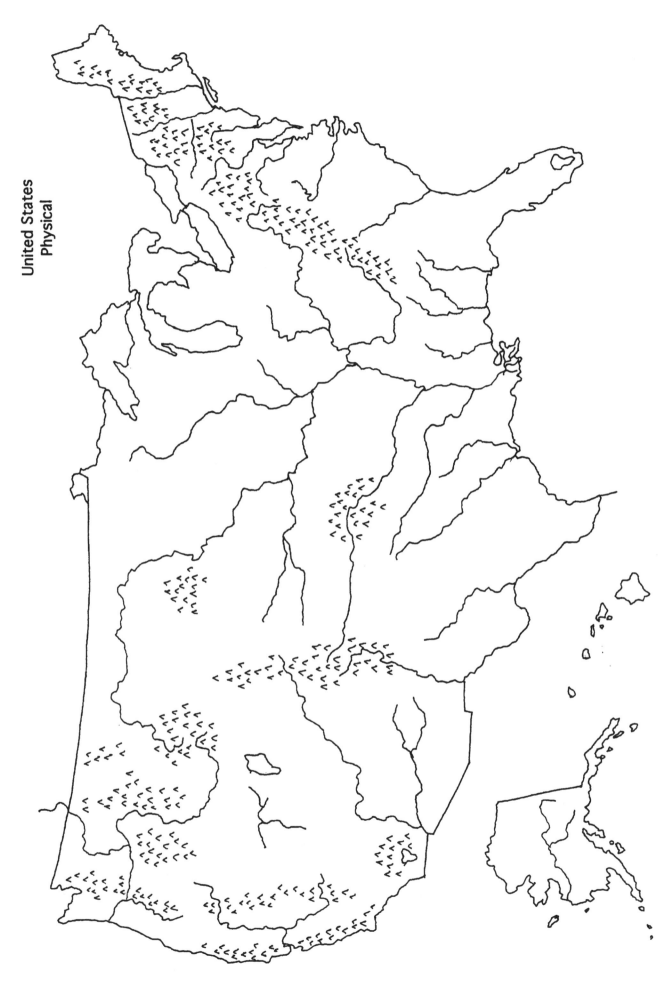

United States
Physical

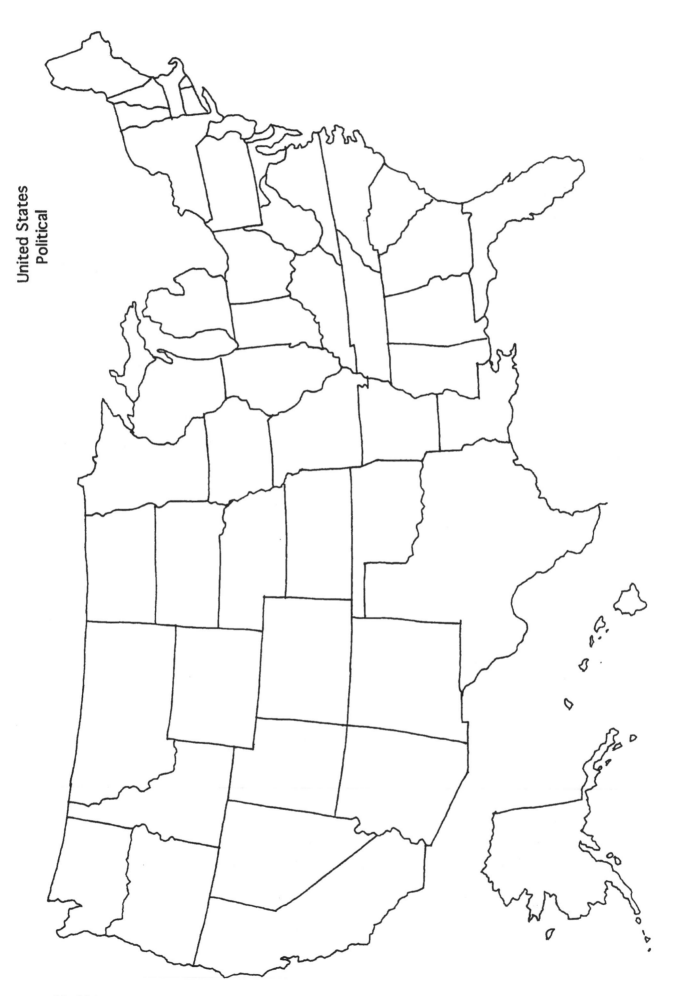

United States
Political

The Northeast and
Mid-Atlantic States

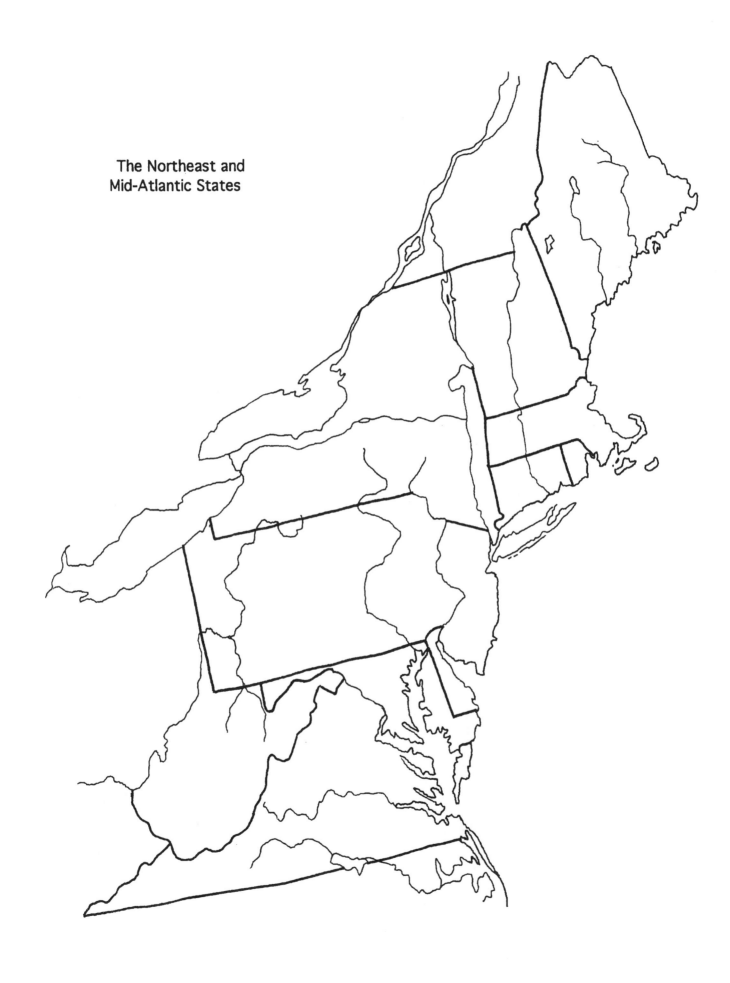

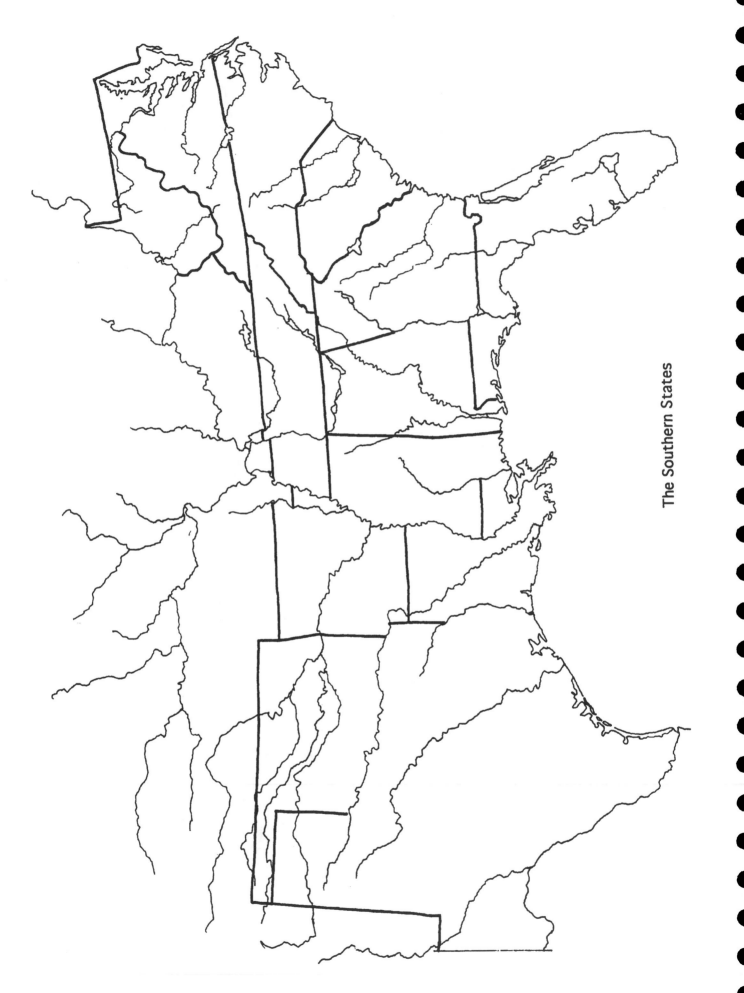

The Southern States

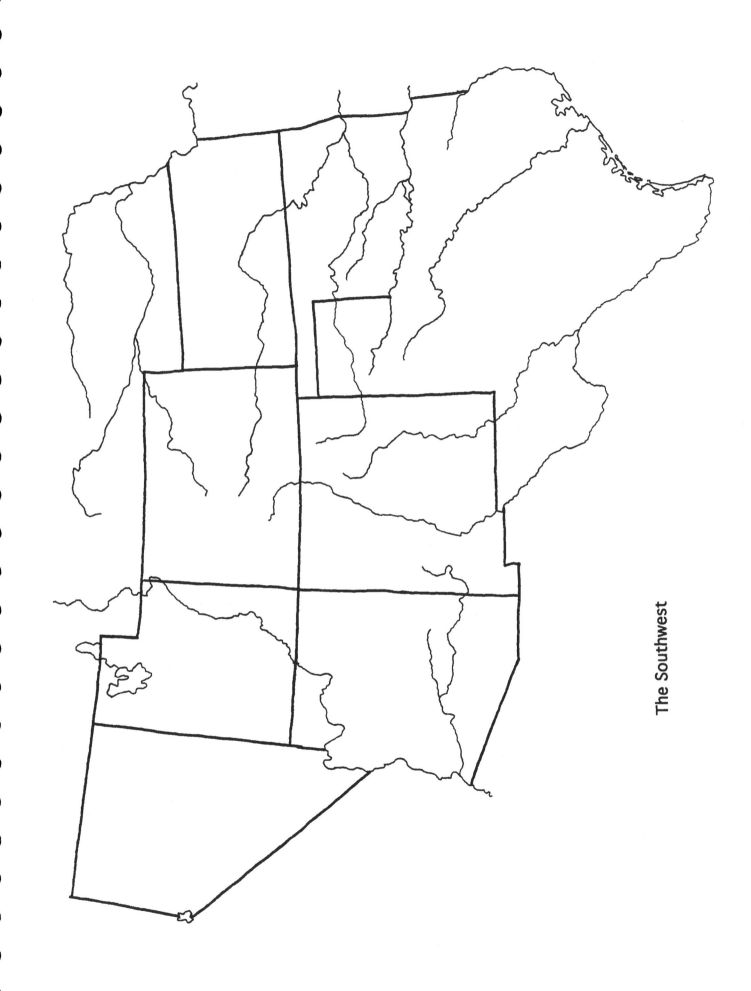

The Southwest

Great Lakes States

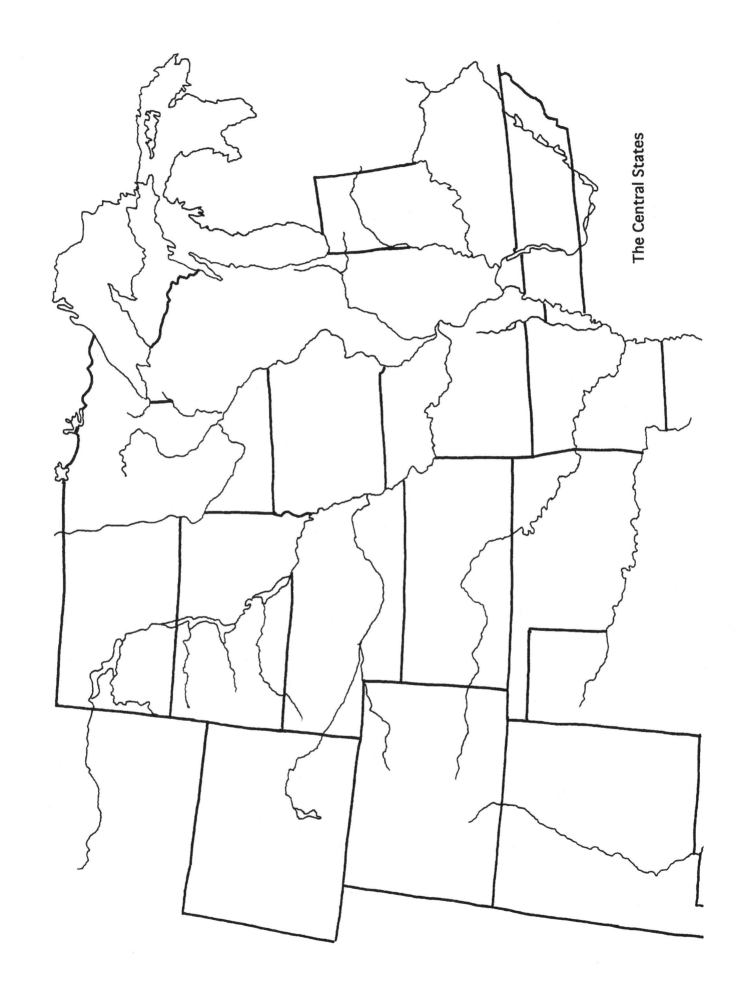

The Central States

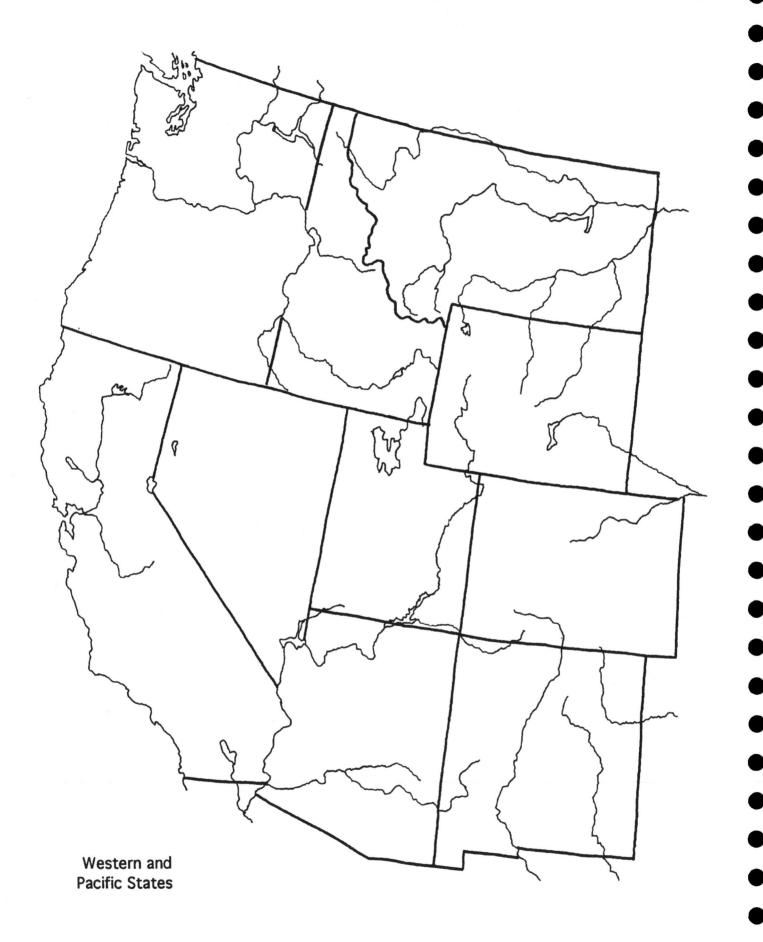

Western and
Pacific States

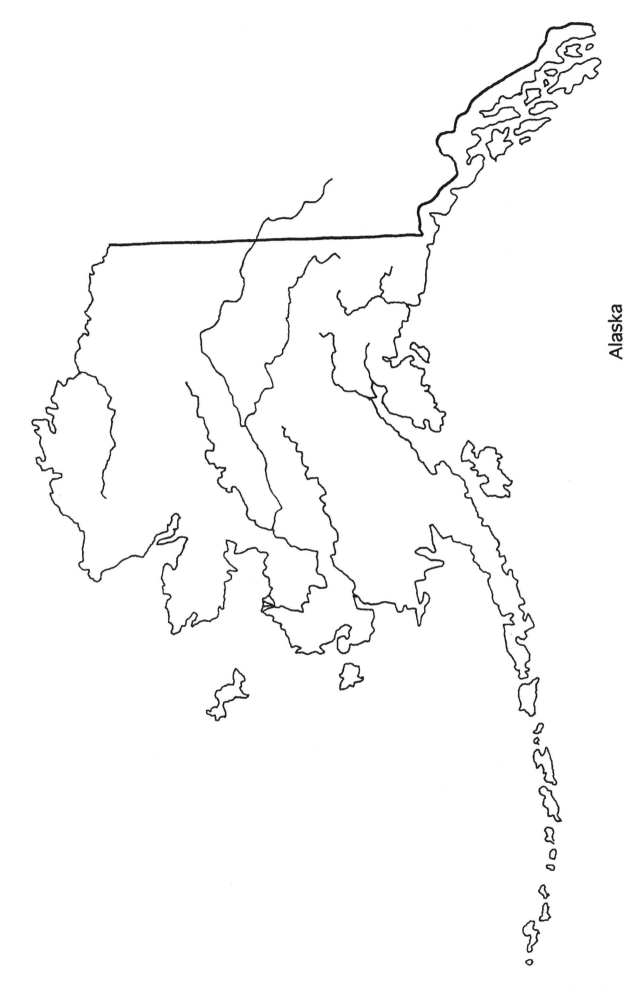

Alaska

Canada

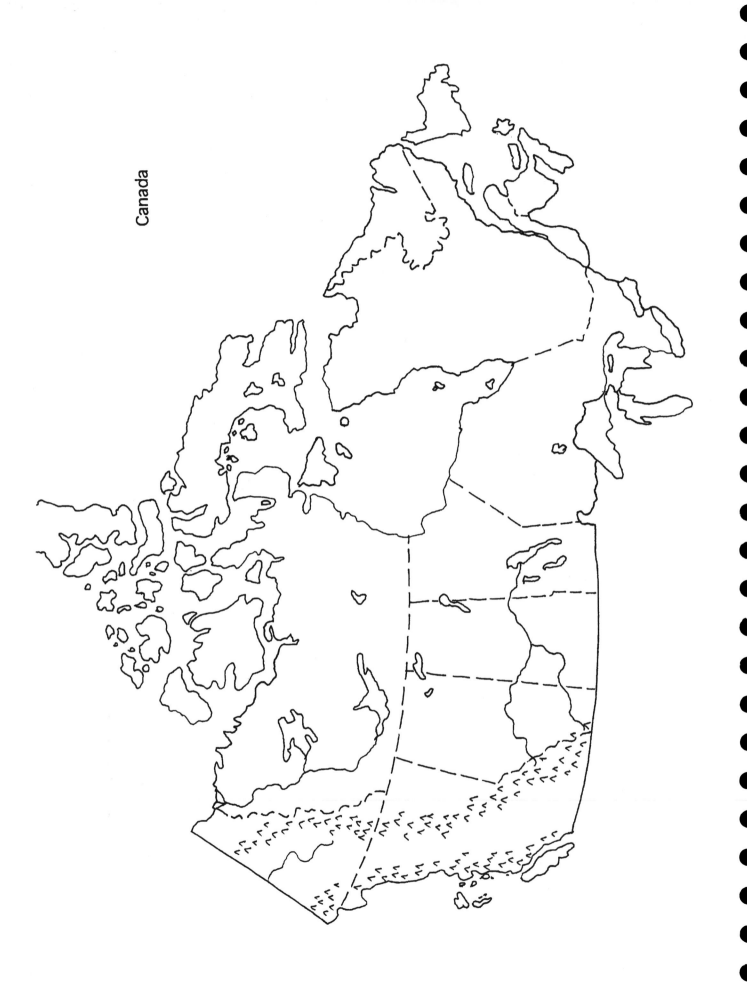

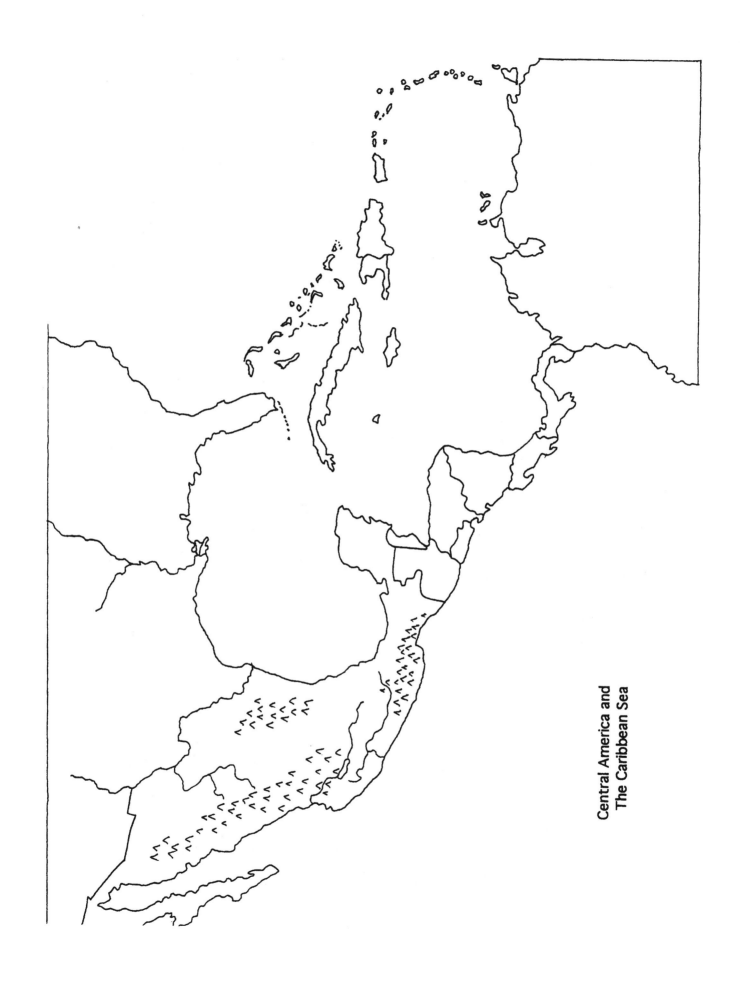

Central America and
The Caribbean Sea

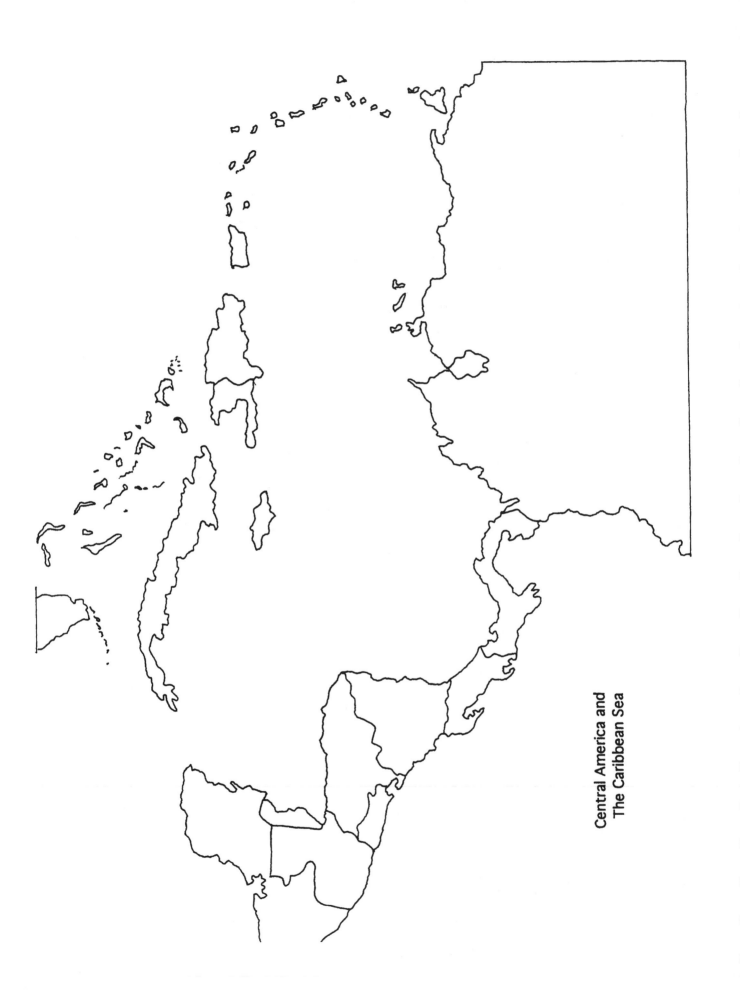

Central America and
The Caribbean Sea

South America

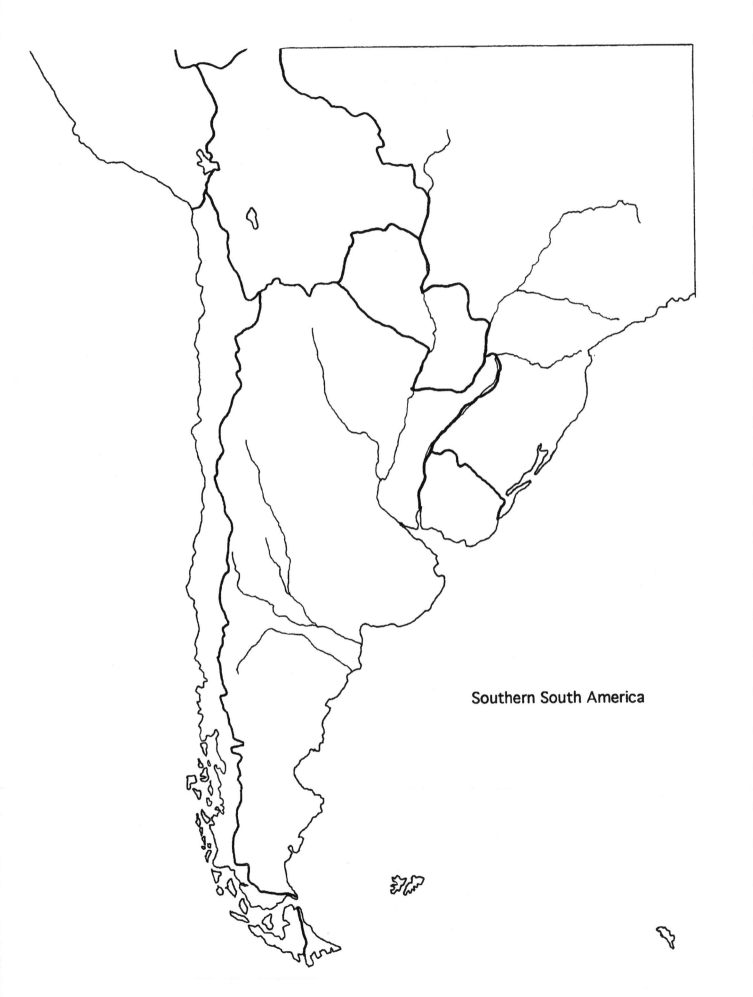

Southern South America

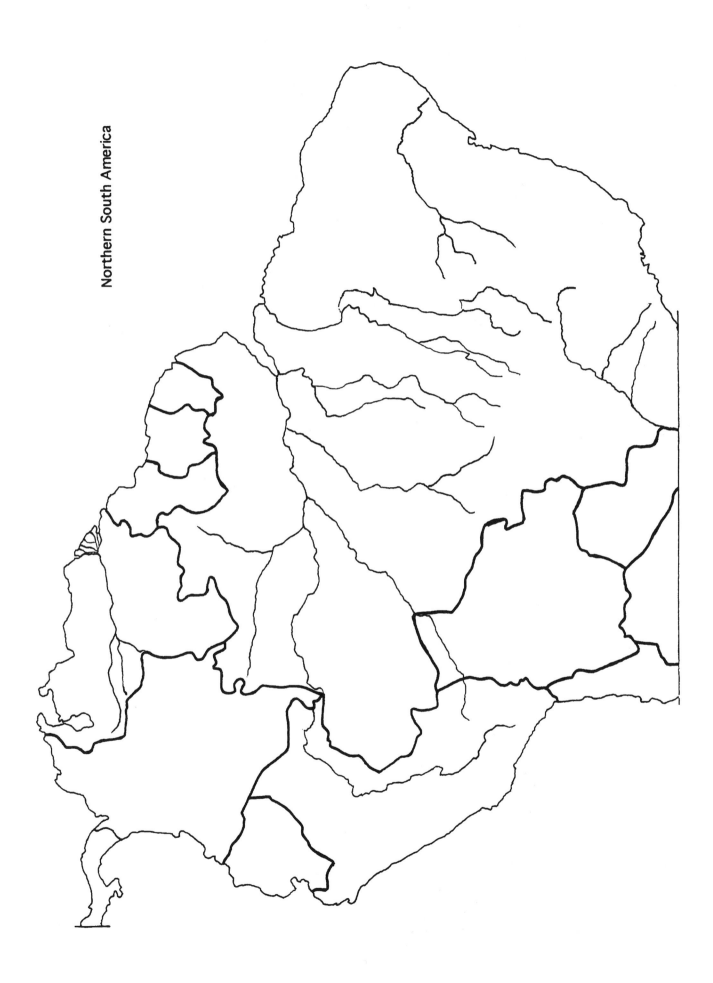

Northern South America

Brazil

Africa

Southern Africa

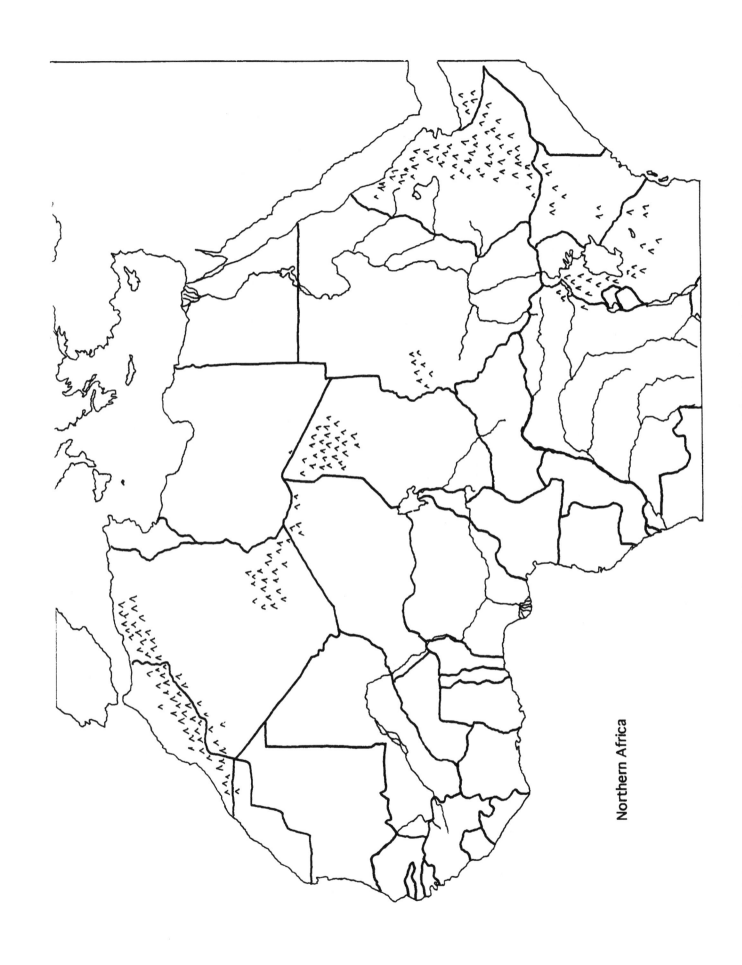

Northern Africa

Europe
Physical

Europe
Political

Central Europe

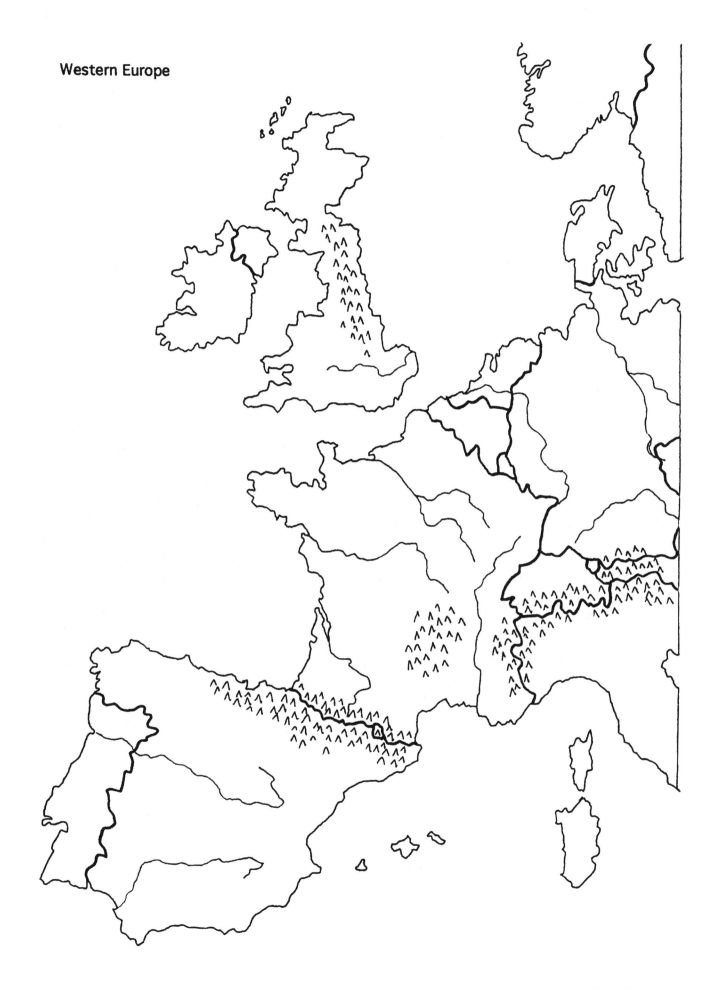

Western Europe

Northern Europe

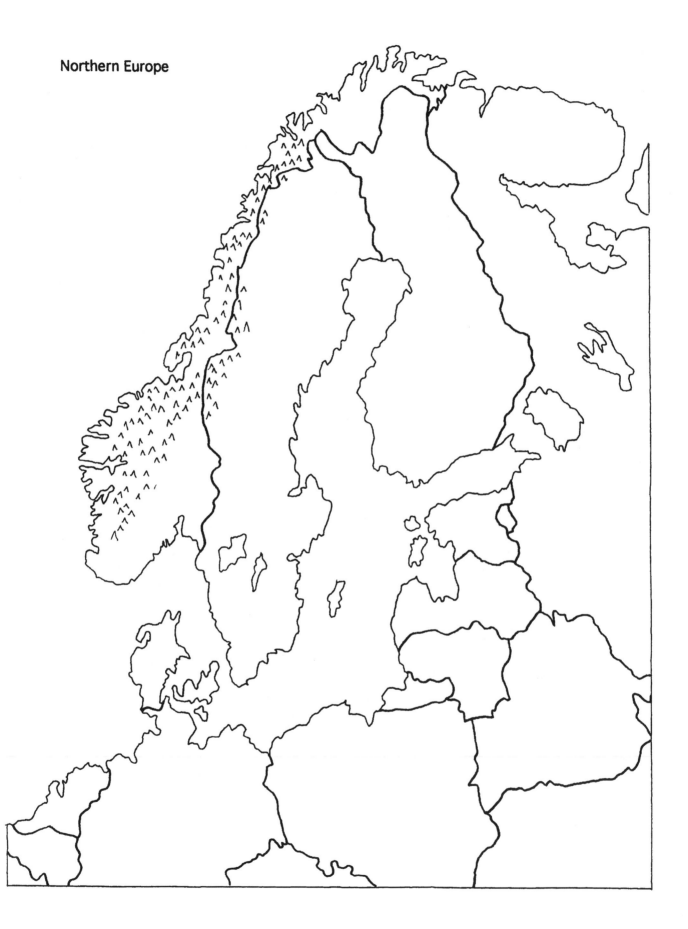

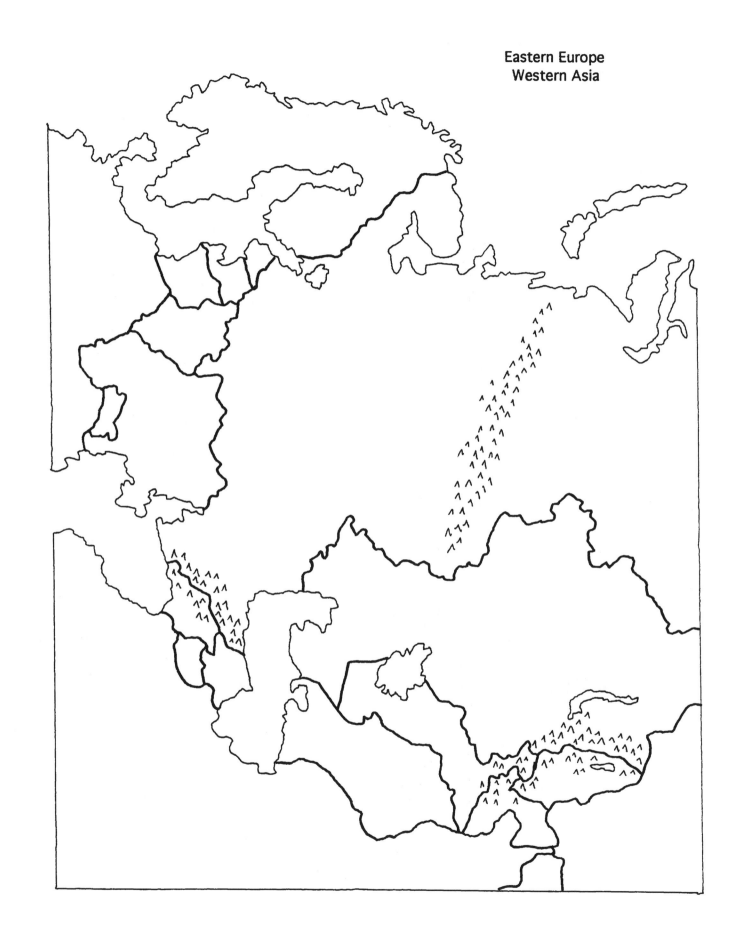

Eastern Europe
Western Asia

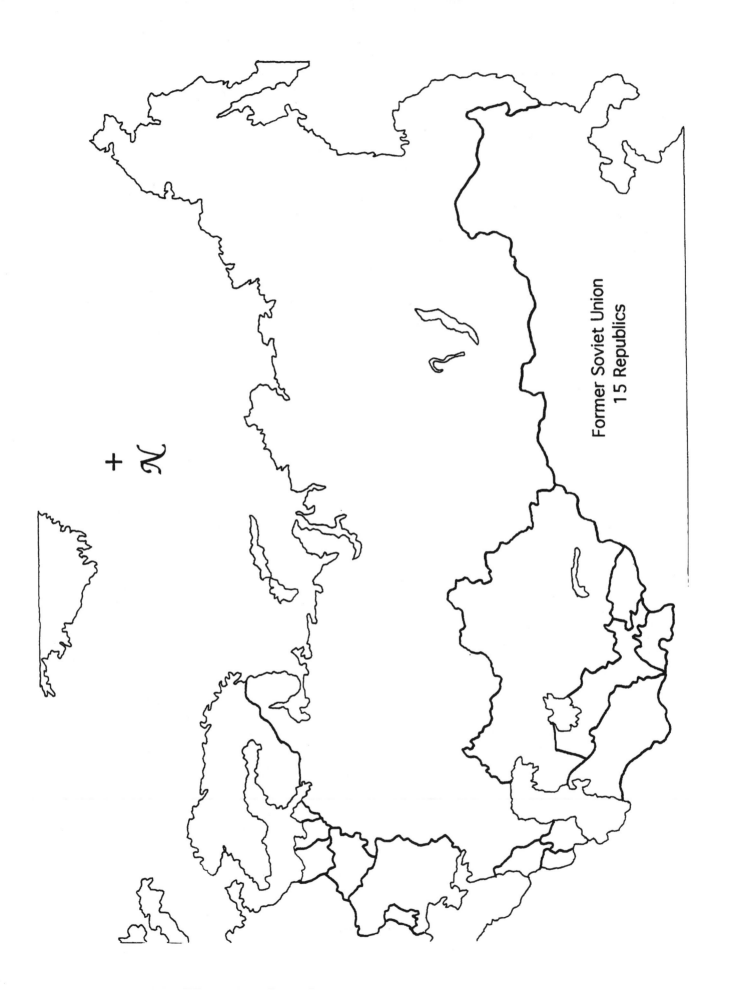

Former Soviet Union
15 Republics

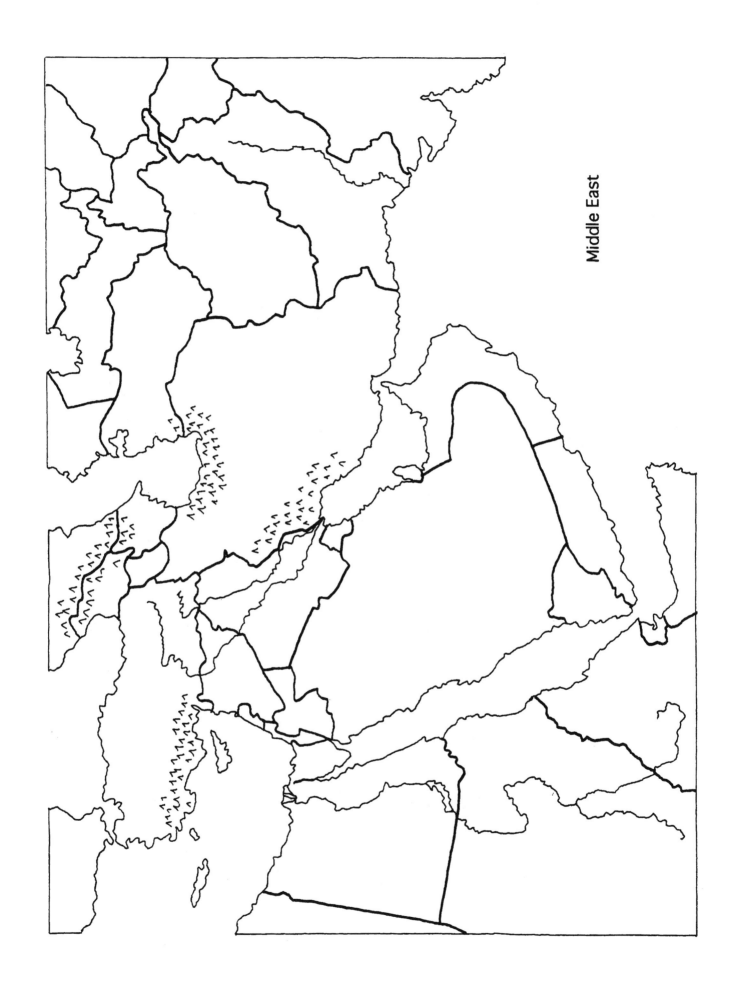

Middle East

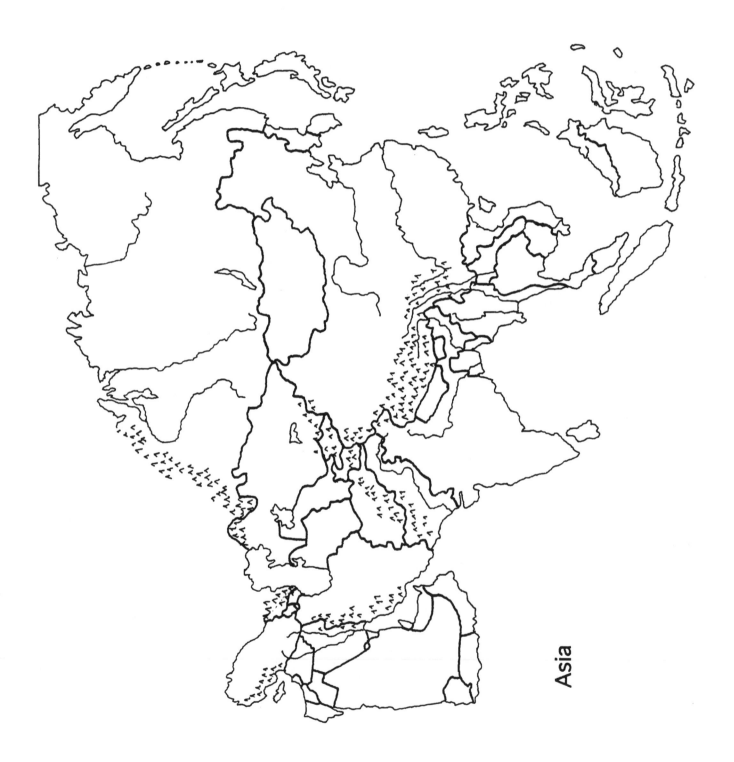

Asia

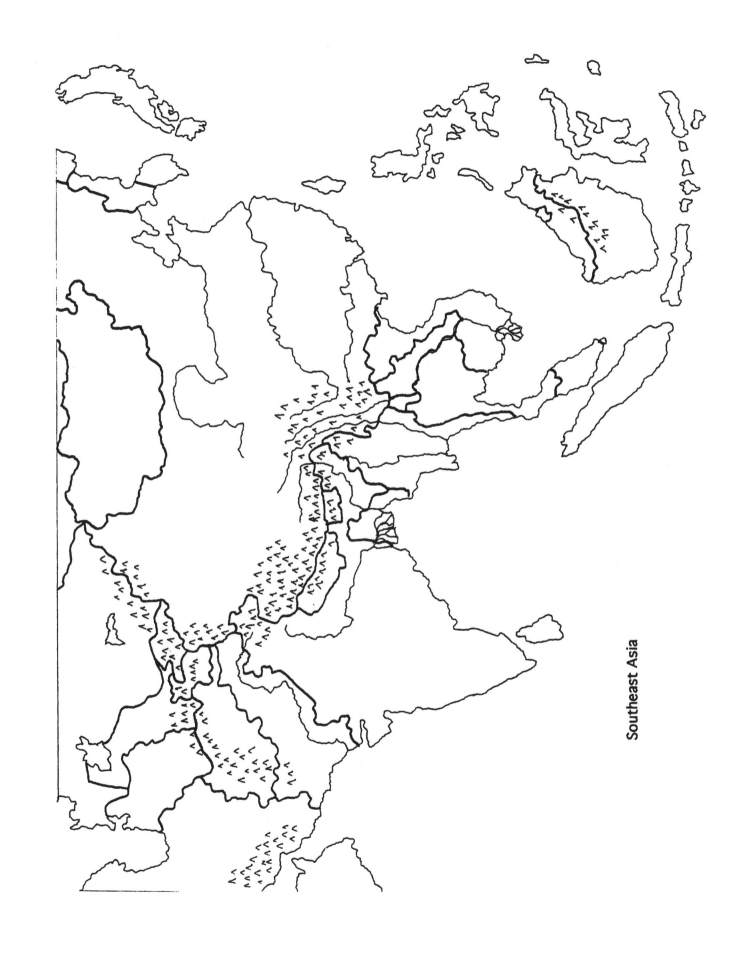

Southeast Asia

Japan

New Zealand

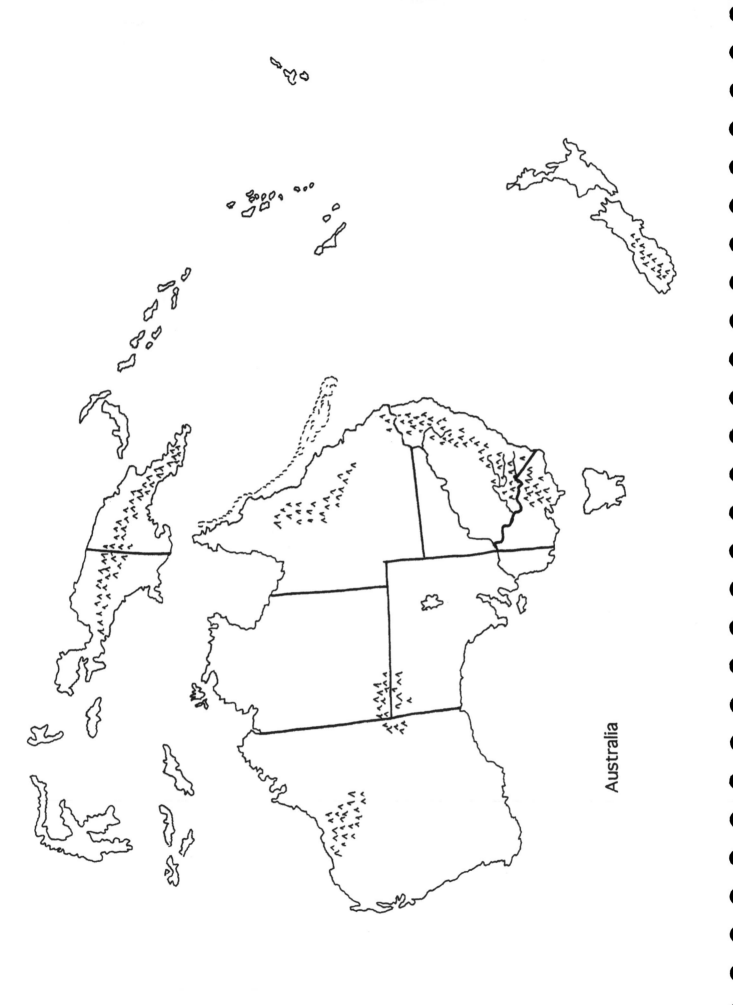

Australia

Index

Notes

Notes

Notes

Notes

Notes

Involve all your students' intelligences

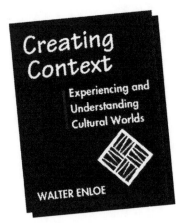

CREATING CONTEXT
Experiencing and Understanding Cultural Worlds

by Walter Enloe
Teacher's resource

Look to Creating Context for a collection of first-rate examples of immersion experiences developed by creative educators. You'll find more than 14 encompassing experiences that involve students from around the world.

224 pages, 7" x 10", softbound.
1066-W . . . $29

Teach how our global community interconnects . . .

MANY PEOPLE, MANY WAYS
Understanding Cultures around the World

compiled by Chris Brewer and Linda Grinde
Grades 4–8

Explore the patterns of life found in nine different areas of the world. Students gain insights into other ways of living by playing games and doing other activities that develop research and critical thinking skills.
  Bring world images to your classroom with 9 full-color posters.

256-page book, 8 1/2" x 11", softbound.
9 full-color, 16" x 22" posters.
1055-W . . . $69

CALL, WRITE, OR FAX FOR YOUR FREE CATALOG!